# The Agricultural Revolution 1750-1880

J D Chambers
G E Mingay

B T Batsford Ltd
London

First Published 1966
*Reprinted* 1966, 1968, 1970, 1975, 1978, 1982, 1984

Printed and bound in Great Britain by
Billing & Sons Limited, Guildford, London and Worcester
for the publishers
B T BATSFORD LTD
4 Fitzhardinge Street, London W1H 0AH

ISBN 0 7134 1358 1

# THE AGRICULTURAL REVOLUTION
## 1750–1880

# Preface

In writing this book we have had in mind primarily the needs and interests of students in Universities and Colleges of Education, while we have tried also to be not unmindful of those of their teachers. If they find, as we hope, this short survey of what is in truth a wide-ranging, geographically diverse and highly complex subject to be not only useful, but stimulating and valuable in a broader sense, we shall think to have succeeded.

Although the agricultural changes of the period 1750 to 1880 remain of prime importance in modern history, there has been no recent book which attempts to draw together in a brief compass all the technical, economic and social aspects of the development. The one general book of this kind, Lord Ernle's *English Farming Past and Present*, first appeared in 1912 and as a text must be considered seriously out of date. Since Ernle wrote, our understanding of the changes of the period has indeed been greatly expanded by research and the discovery of new sources, but for the most part the fruits of this extended knowledge have been presented in learned monographs and articles too numerous, detailed or inaccessible for the non-specialist reader to consult. Of course, in a volume of this size it has proved impossible for us to do full justice to the richness and variety of the material now available on the Agricultural Revolution; but we have tried to incorporate in a fairly simple and straightforward treatment what seem to us the most valuable of recent findings.

In addition to providing an up-to-date summary of the period, we have attempted in some degree also to relate the changes of the period 1750 to 1880 to the broader context of our agricultural development since the Middle Ages, and in particular to point out the unique significance for agriculture—as for the economy at large—of fluctuations in the growth of population. We are not unaware that the use of the word 'revolution' to describe a process of economic change may come too easily to the pen of the economic historian. There is a danger that it may become smooth and slippery and may lose its value in the interchange of ideas in which historians are engaged. In the authors' view, however, the changes which they describe were on such a scale and of such a character as to be justly called a revolution, since they involved a social as well as an economic transformation and were related, in their later phases, to the scientific revolution which is inherent in the transition to the modern world. They had, of course, their roots deep in the past, and they represented an extension of the best existing practices as well as a rejection of others that were already proving obsolete. Their development in the period described was a response to new needs, especially to the needs of a population that, by the last quarter of the eighteenth century, had achieved a modern rate of growth; and their continuous progress during the period of industrialization that followed was a necessary condition for the sustained economic growth that marks this critical period.

For the earlier changes on which the Agricultural Revolution was based, the authors have been able to consult some of the unpublished as well as the published works of two distinguished historians in this field, Dr Joan Thirsk and Dr E. J. Kerridge, a privilege which we wish to acknowledge gratefully here as well as in the appropriate place in the text. They are also under a heavy obligation to Dr Eric Jones, who has read the entire manuscript and has made many useful suggestions; and we are grateful, too, to Dr T. C. Smout and Mrs Rosalind Mitchinson for valuable assistance at various points. As for the division of labour between the authors, their collaboration has been close and constant, and extends over more years than have gone to the production of this book; but in its present form Dr Mingay has the greater responsibility, both for the writing and the research on which it is based.

August, 1965

J. D. CHAMBERS
G. E. MINGAY

So far as possible we have avoided abbreviations in the footnotes, but the following explanation may be helpful:

| | |
|---|---|
| *Ag.Hist.* | *Agricultural History* |
| *Ag.Hist. Rev.* | *Agricultural History Review* |
| *Econ.Hist. Rev.* | *Economic History Review* |
| *J.R.A.S.E.* | *Journal of the Royal Agricultural Society of England* |

# Contents

# Introduction: The Agricultural Revolution in perspective

By the nature of the operations it describes, agricultural history must be distinguished from the history of industry. Agricultural history is concerned with the management of the soil, that is to say with the way farmers throughout the ages have grown their crops while at the same time trying to maintain the fertility of the soil which made those crops possible. The fertility of the soil is the main constituent of the farmer's capital. In yielding a crop the soil gives up something of itself, and unless this is replaced, the farmer's capital, as well as his income, will suffer. Generally speaking, he gets out what he puts in and no more, except where, by a skilful rotation of crops, he can exploit the chemistry of plant growth and so arrange that the plants themselves will leave behind more than they have taken out. And, by the same rule, he can put in only what he has taken out, unless he has access to artificial manures and knows how to use them.

English farmers also had to face the problem of English weather. Fickle and treacherous, it was infinitely co-operative if caught at the right time; and there was a right time for sowing as well as for reaping. Heavy clay soils are more difficult to work than light sandy soils; and when implements were heavy and cumbersome, there were times when cold clay soils could not be worked at all. They might rear sheep or cattle, but they could break the hearts of men, and in a succession of wet seasons, reduce them and their stock to death by starvation. Historians may quarrel as to the meaning of the marks left on the landscape by the plough, but to the farmer they are the signs of the curse of Adam, the need to earn his bread in the sweat of his brow; and when the soil is not only wet but 'clarty', when it falls behind the plough like slices of putty and sets like slabs of concrete, he may sweat in vain. Fitzherbert was not being whimsical when he said, on the subject of when to sow, 'Go uppon the lande that is plowed, and if it synge or crye or make any noyse under thy fete then it is too wet to sow. And if it make no noyse and wyll bear thy horses, thanne sowe in the name of godd.'

This complex relation between the farmer and his main source of capital must be taken into account in any attempt to describe the transition from traditional husbandry practices to modern scientific agriculture and 'high farming'. That is to say, it is no explanation of the Agricultural Revolution to regard it as simply the rural counter-part of the transition to industrialism, the agrarian side of the medal of the Industrial Revolution. It is only necessary to refer to the divergence in the parts played by technological innovation in agricultural and industrial change to know how far this is from the truth. Increased output in agriculture was not, as in industry, to any significant extent the result of mechanical innovation. It is a truism (but one often overlooked) that even in industry machines did not usually take the place of human muscles, except perhaps in the textile industries; they did, however, add enormously to the strength and variety of effort of which human muscles were capable; and they also created an entirely new class of skilled engineers and artisans to produce and maintain them. In the case of agriculture, the various improvements on traditional machinery made an important but still only a marginal contribution to increased output, and the part played by new inventions was almost negligible until towards the end of the period with which we are here concerned. Even the drill, described by Jethro Tull in *The Horse Hoeing Husbandry* in 1733, and with a long history before that, was not generally used for sowing corn before well into the nineteenth century, and on the farms of East Norfolk, the homeland of Townshend and Coke, there was scarcely a drill, horse hoe or horse rake, according to William Marshall, in the last decade of the eighteenth century.[1] As for the manual skills involved, far from being those called into being by the products of the new engineering shops, they were those of the traditional husbandman and stockman and shepherd, working for the most part with traditional tools fashioned in the village workshops. Apart from improvements in ploughs and harrows, the main departure was in threshing, where horse-drawn machinery made its appearance in Scotland in 1780 and was slowly followed by the steam thresher; but the flail continued to be used on the majority of English farms until the 1830s. In harvesting, for an even longer period the 6,000,000 acres under grain had to be mown and

[1] G. E. Fussell, *The Farmer's Tools* (1952) p. 106, quoting William Marshall, *The Rural Economy of Norfolk* (1787) I, p. 59.

reaped by hand, i.e. by scythe and sickle and hand-made straw bands, and the efforts of the women reapers of Scotland, working with the sickle at the end of the eighteenth century, were said to be no more productive in a given period of time than those of the daughters of Tubal Cain.[1]

It is not therefore on grounds of technological innovation that English agriculture can be said to have experienced a revolution. Except for an eddy here and there, the 'wave of gadgets' that is said to have swept over England passed it by until well into the nineteenth century. Looked at from this angle, its mode of expansion corresponds rather with that of the domestic industries that could increase production only by reorganization or by an enlargement of the number of productive units working with traditional tools than with the new factory industries. In terms of actual achievement, however, the increased output of the new agriculture can be shown to belong to an entirely different order of magnitude from anything of which the domestic industries were capable. As early as 1800, and well before the Agricultural Revolution had run its course, British farmers and landlords had accomplished the feat of releasing the latent powers of the soil on a scale that was new in human history; but it was not accomplished by means of a mechanical revolution.

It should also be unnecessary to say that it was done without a ruthless reduction of the rural population as a prelude to the formation of an industrial proletariat. Mental habits die hard, and it is still sometimes thought—and taught—that as a result of enclosure and the introduction of the turnip (one of the most labour-consuming of all the crops in the farmer's calendar) agricultural output rose while the labour force fell—or, as some would say, 'fled'. Agricultural output certainly rose but there is no reason to think that the labour force engaged in agricultural operations fell;[2] the contribution of the Agricultural Revolution was not to release labour for industry, but to make possible a greater output without making a correspondingly larger demand upon the available labour supply. Between 1701 and 1801 the population of the United Kingdom grew from 9·4 millions to 15·9 millions or by 70 per cent. If we discount the supply of foreign grain that found its devious way into our ports to meet the problem of dearth as a result of bad harvests during the war years, it will be seen

---

[1] Fussell, *op. cit.* p. 122.

[2] Contemporary estimates would suggest the contrary. See P. Mathias: 'The Social Structure in the Eighteenth Century: a calculation by Joseph Massie' *Econ., Hist. Rev.* 2nd ser. X (1957–8) pp. 44–5—where the numbers of families engaged in agriculture appear to have risen though the proportion of such families in the total population fell. See also p. 208 below.

that about 6·5 million more people were being supported by the produce of British and Irish agriculture in 1801 than a century earlier.

Expansion of output on this scale without resort to large-scale mechanization or to the exploitation of new virgin land presents a paradox of economic growth that has not been adequately explored, and it is only as a result of the researches of a new generation of agrarian historians that an explanation can be attempted. To refer to enclosure and the taking in of common lands; or to cite the achievements of the somewhat diminished figures of Tull and Townshend and Bakewell will not provide the complete answer. We have to go back to an earlier age, when the traditional use of land was slowly changed so that the fertility of the soil would be increased as a result of a strengthening of animal husbandry with the consequent increase in the supply of animal manure. It begins with the development of convertible agriculture, involving the alternation of arable and grass in place of the ancient division of the cultivated area between permanent arable and permanent grass which tended to undermine the fertility of both.[1] Alternate agriculture implied the practice of arable farming for fodder crops, i.e. the laying down to grass of parts of the arable in temporary leys and the sowing of legumes such as clover, sainfoin or lucerne which added to soil fertility while yielding heavy crops of hay. Hence subsequent arable crops had the double advantage of an increased supply of animal manure and an enhancement of natural fertility through the chemical action of the fodder plants themselves. When to these was added, in the second half of the seventeenth century, the cultivation of the turnip as a regular field crop, calling for heavy manuring and meticulous weeding, the foundation was laid for a new form of land-use, especially adapted to light soils, hitherto suitable only for rough grazing.

This departure from traditional practice marks a new agricultural epoch, and its acceleration in the second half of the eighteenth century in the form of the classical enclosure movement and the first unmistakable steps by the agricultural pioneers towards 'high farming', mark the opening of the Agricultural Revolution just as surely as factory production marks the dawn of a new industrial age. From this time, the Agricultural

[1] This subject is especially well developed in the published and unpublished writings of Dr Joan Thirsk and Dr E. J. R. Kerridge. We have had an opportunity of consulting some of the latter and we would like gratefully to acknowledge their kindness for allowing us to do this. The main characteristics of these important changes have recently been summarized in a valuable article by E. L. Jones: 'Agriculture and Economic Growth in England, 1660–1750: Agricultural Change' *Jour. Econ. Hist.* XXV (1965).

Revolution reveals itself as an indispensable and integral part of the Industrial Revolution, sharing with it the social and scientific attributes that gave the latter its unique character of transition to the modern technological age of mass-production of food, as well as of manufactured goods.

Behind the Agricultural Revolution so defined lies a long period of preparation, the importance and character of which has only recently been realized. Its role in the development of agricultural techniques is epitomized in the fact that whereas in the Middle Ages a harvest yield of four to one of seed was regarded as highly satisfactory, the leading writer on corn production in the eighteenth century took as the basis of his calculations an average of ten to one.[1] The breakthrough in agricultural production had already begun, and it is necessary to consider, at least in outline, the process by which it had occurred if the Agricultural Revolution as we have defined it is to be understood in its proper historical perspective.

The processes of agricultural change are slow and the time span of their germination and maturity is necessarily long. To find the roots of this basic agrarian change, we should probably have to go back to the late Middle Ages when the rigidities imposed by the manorial system were shattered by the demographic catastrophe of the Black Death. Even before this, there were significant pointers to the road which agriculture was to tread. The commercial production of wool and dairy produce by ecclesiastical and lay estate managers and the production of grain for the growing urban markets and even for export, involving the widespread use of marling and the sheep fold, show that the manorial age was not one of entire technical stagnation; and the attention paid to the live-stock industry laid the foundations on which later generations of stock breeders were to build. In his important study of the subject,[2] Mr Trow-Smith shows that cattle and sheep were moved considerable distances to replenish and renew the stock of royal manors; the 'centralized pools of sheep' which were maintained by the great wool-producing estates must have led, he thinks, to the introduction of new types in many districts; and he finds it hard to believe that in 'the late medieval age when the owners of great ecclesiastical estates in particular were striving to improve yields of wool and milk' no importation of continental stock should have been made.

[1] Charles Smith, *Three Tracts on the Corn Trade and Corn Laws* (1766). See also P. Deane and W. A. Cole: *British Economic Growth 1688-1959* (Cambridge 1962) pp. 62-8.
[2] R. Trow-Smith, *History of British Livestock Husbandry to 1700* (1957) pp. 112, 143-4.

Medieval farming, therefore, lacked neither intelligent management nor far-seeing enterprise; but, apart from the burdens and rigidities of the manorial system, it was hamstrung by two technical weaknesses; firstly there was an imbalance between arable and pasture owing to the necessity of increasing the arable, often at the expense of the pasture, in order to provide bread for the rapidly rising population; and secondly, owing to the absence of fodder crops and temporary leys, the supply of animal manure was inadequate for continuous arable cropping broken only by periodical fallow. Animal husbandry was unable to come to the assistance of arable husbandry and there was no known alternative to the traditional rotation of one or two straw crops and a fallow. As a result, in Olaf Stapledon's vivid language, the arable became 'plough sick' and there was a tendency for the already low yields to fall even on the well-managed demesnes of the ecclesiastical estates. As far as the peasant agriculture was concerned, these weaknesses were enhanced by the manorial perquisite of folding the villein's sheep on the lord's demesne, and the small supply of animal manure was further diminished by the practice on the part of the very poor of using it for fuel. The economy of the medieval peasant rested on a razor edge, yet the population continued to grow and the size of holdings began to fall.[1] When distinguished scholars speak of English agrarian conditions sinking to oriental levels, they should refer to the end of the thirteenth, not the eighteenth century.

Only catastrophe could rescue the medieval peasantry and it came in the form of a drastic pruning of peasant households as a result of the Black Death. The vicious circle of fodder shortage leading to soil starvation which hemmed them in was temporarily enlarged; it was not broken. However, conditions were created which made possible the breakthrough without which the economy would have been indefinitely hamstrung by successive crises of rural over-population and falling yields. The condition of deliverance from this cycle of misery seems to have been associated with the rise of yeoman farming for profit on the demesnes and within the village community itself, now released from the strait-jacket of manorial exactions and able to respond by individual effort to the opportunities of the market.

The leasing of demesnes to graziers and capitalist farmers is a commonplace of the text-books; but the emergence of a yeoman class from among

[1] On the Winchester estates the arable land in the hands of peasants fell from 3·3 acres per person in 1248 to 2·5 acres per person '*at the best* in 1311, which in terms of actual cultivation means only 1·66 acres', J. Z. Titow, 'Some Evidence of 13th Century Population Increase' *Econ. Hist. Rev.* 2nd ser. XIV (1961–2) p. 223.

the village population, and the tracing of their expanding substance and their changing husbandry practices through the analysis of the inventories of their possessions at death, is one of the outstanding achievements of post-war agrarian historians. It begins with the work of Dr W. G. Hoskins who, in his articles on *The Leicestershire Farmer*[1] provides the first intimation of the extent to which convertible cultivation in the form of periodical leys had broken into the old routine of permanent arable and permanent grass; and in his study of a single Leicestershire village, Wigston Magna,[2] he explains in personal terms which only a local historian working in minute detail on a small canvas can employ, the process of social differentiation which provided the favourable institutional background for this important change in farming practice. It involved, he says, the emergence of 'what may be called a peasant aristocracy . . . or a class of capitalist peasants . . . the tougher peasants, like the Randulls for example, emerge from the struggle, having acquired the lands of their weaker fellows like the Swetings and the Redleys. . . . One can trace this strengthening and expansion of certain old peasant families over and over again in the Leicestershire village records between the late fourteenth century and the early sixteenth century; one finds as a result that by 1525 there are, in scores of villages, one, two, or three leading yeoman families who own quite a disproportionate share of the personal estate or movable goods.'

Side by side with these, there were of course the more familiar flock-masters on the demesnes, such as John Tame of Fairford who built up a large wool manufactory in Cirencester and supplied it from his own pastures round Fairford which he used as a collecting centre for wool. 'His achievements were replicated by a host of others who lie enshrined in the great churches of the Cotswolds, East Anglia, the Welsh Marches. . . . England, alone in Britain, was in the fifteenth century passing out of an era of monastic and baronial sheep-farming into an area of yeomanly and industrial sheep-farming.'[3] Not only manorial demesnes, but the decayed arable fields of whole villages were taken in, with the consequences that sheep were said to be eating up men. In the next century, however, when the rise of population raised the price of food, and the

[1] W. G. Hoskins: 'The Leicestershire Farmer in the Sixteenth Century', first published in 'Studies in Leicestershire Agrarian History' *Transactions of the Leicestershire Arch. Soc.* (1948), and 'The Leicestershire Farmer in the Seventeenth Century' *Agricultural History* XXV (1951).

[2] *The Midland Peasant: The Economic & Social History of a Leicestershire Village* (1957) pp. 141–2.

[3] Trow-Smith, *op. cit.* p. 148.

balance of profit swung once more in favour of the producer of corn and meat for the expanding urban markets, there was a class of rural entrepreneurs, many of them now in the ranks of the gentry, with the capital and experience to take advantage of it. Landlords whose tenures were sufficiently flexible or who had other advantages (e.g. especially if they held land from the Crown) added the spur of rising rents, and the process of differentiating in favour of the abler and larger farmers by means of consolidation and enclosure for increased arable production hastened the erosion of the peasantry and of the open fields and commons as a feature of the countryside, especially in the areas where animal husbandry could be developed side by side with arable husbandry to the benefit of both. Farmers who owned the land they occupied found themselves on a rising tide of prosperity, and substantial tenants on long leases enjoyed similar advantages and were awarded the same proud title of yeoman farmer. The houses of the yeoman farmers began to spring up alongside those of the gentry, and provided a characteristic contribution to the 'rebuilding of England' to which students of the English countryside have drawn attention.[1]

The response which the rising class of agricultural entrepreneurs made to the new stimulus of the market took various forms, most of them aiming at the increase of fodder supply for the winter feeding of stock. Perhaps the most important was the more systematic use of temporary leys alternating with arable cultivation: and when this was accompanied by the use of legumes or artificial grasses, and particularly clover, a new and important phase of improved agriculture had dawned.

The use of clover seems to have spread from Italy to Holland in the course of the sixteenth century, and by 1620 clover seed was regularly exported to England;[2] but it is not improbable that the value of clover had been accidentally discovered by observant farmers as the result of hard grazing of land that had been improved by dressings of chalk or lime, a system of farming that would induce the voluntary entry of indigenous white clover. Mr G. E. Fussell has recorded an example of clover being discovered in this way by a Somerset farmer in the eighteenth century—in this case red clover—in land that had previously been marled, and it is difficult to think that this was a unique occurrence, however rarely it was recorded. In the early sixteenth century, the best farmers, he thinks, would use the finest seeds from their best hay 'which included

[1] W. G. Hoskins, 'The Rebuilding of Rural England, 1570–1640' *Past and Present* IV (1954).

[2] Slicher Van Bath, *The Agrarian History of Western Europe 500–1850* (1963) p. 279.

clauer (clover) grasse or the grasse honeysuckle', but by the 1650s clover seed could readily be obtained 'at the shop of James Long at the Baye of Billingsgate.' Andrew Yarranton, writing in 1663, thought it was grown in most counties and he believed that six acres of clover were equal to 30 acres of natural grass, not an over-generous estimate, says Mr Fussell, in view of the poor quality of usual wild grass pasture.[1]

By the end of the sixteenth century, the practice of convertible husbandry was making inroads among the strips of the open field farmers, and there is ample evidence that the area of cultivation was being extended by taking in land from the waste[2] which enabled the open field farmer, with or without leys in the open field, to carry a larger head of stock and so to enjoy an increased manure supply. Other important measures were the construction of water meadows, and the folding of sheep on the arable after feeding on the succulent new grass before closing it for meadow.[3] Finally, the turnip, which had been a garden crop in Elizabeth's time, had become a regular field crop in Suffolk by the middle of the seventeenth century,[4] and Sir Robert Weston, after his period of self-imposed exile in Flanders, was advocating its cultivation, and also giving instructions on how to prepare a five-year clover ley. The age-old problem of finding winter feed for stock was on the way to solution, and farmers could now save their best animals for selective breeding, especially when they were operating, as many of them were, from the secure base of enclosed farms.

It would be surprising if the decisive advantage which farming in severalty gave to the improving farmers was not reflected in a new attitude to the laws relating to enclosure. Sir Walter Raleigh epitomized it in the debate at the end of the sixteenth century when he said that the law should be so framed to 'let every man use his Ground to that which it is most fit for, and therein use his own Discretion.'[5] But neither the theory nor the

[1] See G. E. Fussell, 'Pioneer Farming in the late Stuart Age' *J.R.A.S.E.* (1940), 'Low Countries' Influence on English Farming' *English Historical Review* LXXIV (1959) pp. 615–17; 'Adventures with Clover' *Agriculture* Oct. 1955.

[2] E.g. in Laxton, Nottinghamshire, 1,500 acres out of a total of 2,800 acres was already in arable or pasture closes by 1635. See C. S. & C. S. Orwin, *The Open Fields* (Oxford, 1938) p. 121.

[3] See especially E. J. Kerridge, 'The Sheepfold in Wiltshire and the Floating of the Water Meadows' *Econ. Hist. Rev.* 2nd ser. VI (1953–4).

[4] E. J. Kerridge, 'Turnip Husbandry in High Suffolk' *Econ. Hist. Rev.* 2nd ser. VIII (1955–6) p. 390; see also E. L. Jones, *op. cit.*

[5] See M. Beresford, 'Habitation Versus Improvement' in *Essays in the Economic and Social History of Tudor and Stuart England*, ed. F. J. Fisher (Cambridge, 1961) p. 45.

practice of 'laissez-faire' in land-use could be expected to pass without challenge: it was noted that 'the ears of sheep masters doe hang about the doors' of the House, and the memory of their depopulations was too recent and too bitter to be easily forgotten. Indeed, it was being fanned into new life in some parts of the country; and the procedure for giving legal sanction to the agreements to enclose for increased production of corn and meat and dairy produce found ways of discreetly by-passing the open forum of Parliament by taking the form of Decrees in the Royal Courts of Chancery and Exchequer. In this age of parliamentary independence, the ancient prerogative courts had their uses, and in their unobtrusive legalization of private agreements to enclose they anticipated the main characteristics of the Enclosure Acts of a later date, even to the point of employing enclosure commissioners and surveyors. The history of enclosure by formal legal instrument, as Professor Beresford has shown, does not begin with Hanoverian Acts of Parliament; it lies hidden in the unnumbered Enrolled Decrees in the Public Record Office, in the private agreements in Estate Records, bishops' courts and parish glebe terriers.

In some areas it could be said that the changes had brought about a limited and localized agricultural revolution by the middle of the seventeenth century. In the neighbourhood of London, for instance, the expansion of the market under the stimulus of population growth had created a market garden industry which by 1650 led to 'a minor revolution in the ordinary citizen's diet'. There had been a trickle of produce—cabbages, cauliflowers, turnips, carrots, parsnips and fruit—from the gardens and orchards of the gentry of Essex and Kent from the early years of Elizabeth's reign; by 1635 the open field farmers were taking part and helping to raise the trickle to a stream of 'some proportions'.[1] There were husbandmen and yeomen who had risen to be gentlemen farmers by fattening cattle for Smithfield market; there was specialization arising from the selective pressure of the London market and regions were becoming interdependent through concentration on particular products and breeds of cattle.

It can also be shown, however, that important as these advances were, supply could not keep pace with demand, and, in the case of London, the City's appetite grew faster than the country's ability to satisfy it. That is not surprising in view of the reported growth of population from about 50,000 in 1500 to 500,000 in 1700; and since there is no evidence that foreign supplies were any more plentiful than before, there were com-

[1] See F. J. Fisher, 'The Development of the London Food Market 1540–1640' *Econ. Hist. Rev.* V (1934–5) pp. 53–4.

plaints that the needs of London were met by 'pinching the bellies of the poor' in the country. Dr Hoskins has recently shown that between 1520 and 1620, every fourth year, on an average, was marked by deficiency tending to dearth; and 'in a country in which between one-half and two-thirds of the population were wage-earners, and a considerable proportion of the remainder subsistence farmers; in which about one-third of the population lived below the poverty-line and another third lived on or barely above it; in which the working class spent fully 80 to 90 per cent of their income upon food and drink,'[1] a deficiency of harvest leading to dearth—i.e. famine—every four years is a dubious testimonial to the efficiency of agriculture.

Moreover, the areas capable of improvement were limited: the 'Great Level of Fen', an area of 700,000 acres in the counties of Cambridge, Lincoln, Huntingdon, Northampton, Suffolk and Norfolk, still awaited the campaigns of Vermuyden and the Duke of Bedford; and the art of underground drainage of the traditional clay lands was in its infancy. The Royal Forests remained virtually untouched and large tracts of land were allowed to be waste so that, as a writer complained in 1656, 'there are more *waste* lands in *England* than in all Europe besides, considering the quantity of land.' He adds, that 'poore people will cry out against me because I call these waste land: but it's no matter.'[2] The time was not ripe for enclosure by compulsion, but, unless there was a check to population growth, it could not long be delayed.

The obstacles to agricultural growth were thus physical and political as well as technical. The advances of the early pioneers were of great significance, but they were made on too small a scale to meet the demands of the Elizabethan population which was growing at an unusually rapid rate, possibly owing to the diminished virulence of epidemics in the second half of the sixteenth century; and the periodical famines noticed by Dr Hoskins are partly a reflection of the relatively greater speed of population growth compared to that of food supply, each operating under its own complex factors of change. In the next century, the position seems to have been reversed. To quote Professor Fisher again, '... it is perhaps significant that the second quarter of the seventeenth century, when the upward swing of agricultural prices began to flatten out and real wages began to rise, was also a time when disease and emigration were probably

1 See W. G. Hoskins, 'Harvest Fluctuations and English Economic History 1480–1619' *Ag. Hist. Rev.* XII (1964) p. 28.

2 R. Child, 'Large Letter written to Mr. Samuel Hartlib' (1655) quoted Lord Ernle, *English Farming Past and Present* (6th ed. 1961) p. 110.

combining to check the rate of population growth.'[1] The improved
agricultural system was still geared to what was virtually a medieval rate
of population growth, and while that condition obtained, it is not surprising if the advance in technique and the enlargement of the area, e.g.
by drainage, should again enable supply to keep pace with demand when
the rate of population growth slackened. Indeed, from the middle of the
seventeenth century the fear of famine was clearly receding and in its
place there was more often heard the farmers' complaints of the 'evil
consequences of plenty'. The Corn Bounty Acts were passed to meet them
by encouraging exports, and during the lean years of the 1690s England
usually had a surplus for export. Even in the disastrous famine years
1709–10, when English prices rose to a level not surpassed until the last
years of the eighteenth century, England was able to supply the needs of
less favoured countries. 'In 1711, 1713 and again in 1715, though yields
of wheat were light, the existence of substantial exports of grain suggests
that England was free from hunger.'[2]

Freedom from hunger, when population growth was again unmistak-
ably on the upgrade and when, for three years, 'the wheat yields were light':
this in itself is sufficient warrant for the view that agricultural output had
achieved an advance that marks the dawn of a new agricultural epoch.
Farmers had learned how to conserve and add to their capital in the form
of soil fertility; on the light soils at least they had broken through to
increasing returns on agricultural investment. This is an achievement of
supreme importance; but it was a condition rather than a fulfilment of the
promise of agricultural revolution. It corresponds to an achievement on
the side of industrial innovation such as the pound lock which harnessed
the energy of water supply to the tasks of transport and motive power.
Compared to some other industrial innovations of the time—Newcomen's engine and Edward Wright's 'cupola' furnace—agricultural progress lagged behind that of industry in that it owed virtually nothing to
contemporary science. As the writings of Jethro Tull showed, it was
possible for one of its leading figures to do the right thing for entirely the
wrong reasons, and there is no evidence to think that the most enlightened
of his contemporaries could have offered better. They still depended on
trial and error, on the uncanny instinct—'the feel'—of the dedicated
farmer for doing the right thing with soil and plants and animals without
appreciating the nature of the reasoning that lay behind their decisions.

[1] Fisher (ed.) *Essays in the Economic and Social History of Tudor and Stuart England*
p. 3.
[2] T. S. Ashton, *Economic Fluctuations in England 1700–1800* (1959) p. 17.

In this respect agricultural progress continued to lag behind that of industry, especially on the side of animal husbandry, because of the relative backwardness of the biological sciences; and the progress made in this branch of agriculture is all the more remarkable in that it depended on the instinctive insights of men of genius who were able to anticipate the findings of genetical and veterinary science in their breeding practices. They could do this, however, only because the experience of an earlier age had created the tools of the new animal husbandry in the form of new fodder crops and improved localized breeds of animals that centuries of commercial agriculture had slowly evolved. Robert Bakewell and the Colling brothers were related to them in much the same way as Watt and Cort were to Newcomen and Darby, or more appropriately still, as Brindley to the first builders of the pound lock; but historians have no doubt as to which of these pairs of industrial innovators were actual participants in and which precursors of the Industrial Revolution; and the same distinction could appropriately be made in the case of the two sets of agricultural pioneers.

The question now arises as to the nature of this distinction; and the answer would appear to be found first in the scale, and secondly in the methodology, which characterized the progress of innovation in the two periods. In the second period the scale of innovation was geared not to a medieval but to a modern rate of population growth, and thanks to the work of earlier pioneers the productive processes of both agriculture and industry were able to respond to it, according to their respective capacities. In the case of industry the response came through the increase in productive units and the great acceleration of technological innovations; in the case of agriculture, it came through expansion of acreage, reorganization of existing holdings and, especially, the further application of known techniques of soil and animal husbandry for increasing production and productivity to meet the challenge of the new rate of population growth. From less than 0·2 per cent per annum in the middle of the century, population was rising at the rate of 1·4 per cent per annum during the years of war with revolutionary France; and until the war was over, only the capacity of British agriculture to tap the potentialities of the soil stood between the nation and sheer famine.

The second distinctive characteristic of the new era of production was the growing realization of the almost illimitable potentialities of both industry and agriculture under the influence of scientific direction. In this respect agriculture continued to lag behind industry, with unfortunate results on contemporary thinking which remained under the influence of the myth of decreasing returns long after the means lay to hand by which it could be refuted. The importance of phosphates in plant growth had been known from the 1790s but it was not until Justus von Liebig showed the way to the manufacture of superphosphates in 1840 and John Barnet Lawes began to manufacture them in his factory at Deptford in 1843 that the possibilities of scientific agriculture began to be realized. The application of chemistry to the problems of farming, together with the mastery of underground drainage and the introduction of mechanical reaping and other forms of mechanization in the middle decades of the century, marks the fulfilment of the promise of plenty for all, no matter what the pressure of demand may be. It takes us over the threshold into the modern world of scientific technology for the mass-production of food, as well as for manufactured goods, and for that reason may be said to complete the transition which we continue to call the Agricultural Revolution.

*References and suggestions for further reading:*

| | |
|---|---|
| A. and N. L. Clow | *The Chemical Revolution* (1952), Ch. XXI. |
| D. C. Coleman | 'Industrial Growth and Industrial Revolutions', in *Essays in Economic History* (ed. E. M. Carus-Wilson) III (1962). |
| G. E. Fussell | *The Farmer's Tools* (1952), Ch. III–V |
| W. G. Hoskins | *The Midland Peasant: the Economic and Social History of a Leicestershire Village* (1957), Ch. VI, VII. |
| E. L. Jones | 'Agriculture and Economic Growth in England 1660–1750: Agricultural Change' *Journal of Economic History* XXV (1965). |
| E. L. Jones | *Agriculture and Economic Growth in England 1650–1815* (1967). |
| *History of Technology* IV (ed. C. | Singer, E. J. Holmyard, A. R. Hall, and T. I. Williams, 1958), Pt. I. |
| E. Kerridge | *The Agricultural Revolution* (1967). |
| B. H. Slicher van Bath | *The Agrarian History of Western Europe 500–1850* (1963), Pt. 3, Section D. |
| R. Trow-Smith | *History of British Livestock Husbandry to1700* (1957), Ch. 3–4. |
| Charles Wilson | *England's Apprenticeship 1600–1763* (1965), Ch. 2, 7. |

# 1 Agriculture in the Early Eighteenth Century

At the beginning of the eighteenth century land and its cultivation continued its ancient domination of society and the economy. According to the computations of the contemporary statistician Gregory King, landlords' rents and farmers' profits accounted for about a half of the country's incomes;[1] and if we take into account the agricultural labourers, the rural craftsmen such as blacksmiths and wheelwrights, the carriers, brewers and other tradesmen, and the professional men like attorneys and doctors whose incomes came largely from the country folk, then agriculture could be said to support about three-quarters of the population. The cultivated land of England and Wales, including woods, orchards and gardens, covered about 25,000,000 acres according to King, maintained 12,000,000 sheep, 4,500,000 cattle and 2,000,000 pigs, worth together about £15,000,000, and bore crops worth about £9,000,000.

But perhaps we get a clearer picture of the dominating role of agriculture if we consider its various vital functions in the economy. First of all, the land was of course the supplier of food for the people, and of the raw materials which went to make beer, ale, whisky, and gin, clothing, shoes, furniture, candles, soap, starch, knife-handles and glue; and of course the land supplied horses for agriculture, industry and the carrying trade, as well as timber used in houses and ship-building, and in making carts, wagons and coaches. According to King 3,000,000 acres were devoted to woods and coppices, and their produce ranged from great

---

[1] Phyllis Deane and A. W. Cole estimate that agriculture was responsible for 43 per cent of the total national output in 1700: *British Economic Growth 1688–1959* (Cambridge, 1962) p. 78

beams of oak intended for ships of the line to slender hop poles and humble barrel staves, not forgetting the cordwood which was burned to provide the charcoal used in ironworks. The millions of sheep grazing the downs and wolds were valued for their wool as well as their mutton, and hides and tallow were valuable by-products of cattle-raising. Dairying and the keeping of pigs went together because the pigs could be fed on the skimmed milk and whey, and so the dairy farmers produced much pork, bacon and ham. The great majority of all farmers, however, kept a pig or two, if only for their own table, because they were so easily fed on scraps or could forage for themselves in the woods. The keeping of some cattle and sheep was usually associated with arable farming because cattle provided draught animals and food, while sheep provided manure to keep the plough-land in heart, as well as meat and wool. Wheat was the most valuable crop, but it is interesting that King thought that over twice as much barley was grown as wheat, and even more oats than wheat, the oats being used for food in the north and also for keeping the innumerable horses and oxen employed as draught animals. Wheat, barley and rye were often mixed in making the 'maslin' bread (and were sometimes sown together in the fields), and barley was sent to the distilleries and was malted for brewing beer and ale. In turn the waste products of breweries and distilleries were used for fattening pigs, often on quite a large scale, for the London market and other centres of consumption.

English farming was not self-sufficient: a large proportion of the young cattle were brought in for fattening from Wales, Scotland, and later Ireland, and the nearby continental countries, together with some dairy produce, but in the first half of the eighteenth century we had in many years a sizeable export of grain. About 24,000,000 quarters were exported between 1732 and 1766, and the 1730s and 1740s, years of exceptionally good harvests, saw very high exports of wheat, and also of barley, malt and rye. It was estimated that in 1750 nearly a quarter of that year's total wheat crop was exported. This, however, was the last period in which England figured as an important exporter of home-produced foodstuffs, and within two or three decades of 1750 we had become on balance an importing country.

The system of transport depended heavily on water, on coastal shipping and navigable rivers, but also on the 'land carriage' for short-distance hauls and where the cheaper water carriage was not available. The land carriers hauled loads of grain, butter, cheese, fish, timber and wool, as well as the produce of manufacturing industry, ironworks, brick-kilns, quarries and mines, and many of the carters and wagoners were part-time farmers or at least the occupiers of land which they used for pasturing their horses and oxen. The employers in domestic industries relied heavily on the labour of the farming families for spinning and carding, and the making of such articles as lace, gloves, ropes, nets and sacking. Under-employment in the countryside, especially during the winter months, provided a large supply of cheap labour for these tedious and poorly-paid crafts, but in the summer the tables were turned when the good wages, food and drink of harvesting brought such work to a stand-still and even attracted the weavers and framework-knitters away from their machines to the harvest fields.

Finally, an essential function of land was to support a body of land-lords, who together with the urban middle classes, constituted the governing class. Because of their wealth, influence and command of patronage, the few hundred great landlords, whose extensive but scattered estates ran into many thousands of acres, were able to control Parliament and dominate government. The far more numerous class of lesser landlords or gentry owned together about half of the land in the country but were content with a subsidiary role in government. Although the majority of seats in the Commons were occupied by country gentlemen, only a few of the more talented and wealthy among them, gifted politicians like Walpole, managed to attain high office; the generality of squires could not even afford to contest an election, and devoted their energies to the care of their estates and local affairs. As Justices, however, many of them were persons of consequence in their own neighbourhood, having responsibility for maintaining law and order and the power to try offenders at Petty and Quarter Sessions. The importance of the Justices was enhanced by their administrative authority over the poor law, over markets and fairs, weights and measures, roads and bridges and a multitude of other matters—matters affecting much more the day-to-day life of the country than the Acts passed in Parliament.

The great landlords and gentry were primarily a leisured, educated class of rulers and administrators who drew their income from the rents of land and such estate profits as arose from sales of timber and mining royalties. To a lesser extent they also depended on the holding of invest-ments such as mortgages and the Funds, and from the profits of govern-ment offices, pensions and sinecures. Although many of them employed no steward but personally managed their estates, supervising the tenants, keeping the accounts, and dealing with the humdrum questions of leases, tithes and repairs, only the smaller country gentlemen were usually practising farmers. They frequently kept part of their land permanently in their own hands and relied considerably on its profits. The home farms of the larger owners were managed for them by a bailiff, and served principally as a convenient source of food supply for the large families and cohorts of servants who were customarily resident in the great houses.

In England and Wales the farmers proper numbered about 330,000 according to Gregory King. Nearly half of them, 150,000 he estimated, were tenants who farmed the land belonging to the great landlords and gentry. The remaining 180,000 were owner-occupiers, farming their own land, and sometimes renting additional land from a landlord. The vast majority of the independent owners King placed in the category of 'lesser', and the amount of land owned by each of these was probably small, averaging perhaps less than 20 acres. Below the small owners were of course the cottagers and labourers. It is impossible to say exactly how numerous they were, but from King's figures it has been estimated that they were not very much greater in number than the farmers—perhaps only seven labourers to every four occupiers of land. Even in 1831, when farms had grown in size and many small occupiers had disappeared, the proportion of labourers to land-occupiers was still only eleven to four.[1] The picture sometimes presented of English farming, with a select band of large capitalist farmers employing a vast army of landless labourers, is patently a false one. The early eighteenth-century proportion of $1\frac{3}{4}$ labourers to every occupier is less surprising when it is remembered that most English farms at that time were small, and that at least half of them could be worked with the labour of the farmer's family, no hired help being necessary except perhaps at harvest time. Even the larger farms, Arthur Young calculated, required only one hired hand, man, boy or milkmaid, per 40 acres, and in our period any farm of more than 300 acres was considered large. Arable farming required much more labour than did permanent pasture: about two men for every 50 or 60 acres of arable compared with one man for the same area of grass. But labour requirements were complicated by the heavy seasonal pressure in spring and summer and by the farming methods employed: the hoeing of turnips as in the Norfolk system was heavy on manpower, while convertible husbandry in which grass and arable alternated required more labour than land permanently under grass.

At the beginning of the eighteenth century many of the labourers were hired by the year and lived in with the farmer and his family (and were then known as farm servants); otherwise they might still be hired by

[1] J. H. Clapham, *Economic History of Modern Britain* I (1926) p. 114.

the year but live out in cottages as out-servants, or were paid by the day as mere day labourers. Many cottages carried with them the right to use the common for grazing a few beasts and for getting fuel. In open-field districts, and sometimes elsewhere, it was possible for cottagers to have a small plot of land which they might use for raising vegetables and a little corn or for grazing purposes. For a large proportion of the labouring population, therefore, there was the security of the annual hiring or the supplementary income of the commons and perhaps a small holding. As time went by, however, the situation gradually changed. Living-in declined from the later years of the eighteenth century as farmers' incomes rose and with them their social pretensions; and their wives and daughters demanded (with) drawing rooms and dining rooms from which the farm servants were excluded. The rising cost of food also helped to incline the farmers towards this change, although farm servants remained strongly entrenched in the more conservative north and where the large number of animals kept required the constant care of servants on the premises. The cottagers, too, gradually lost their common rights and open-field plots as both commons and open fields were swallowed up into enclosed farms, and they had to rely more and more on their employment on the farms.

Nor were the owner-occupiers and small gentry immune from the effects of change. In the late seventeenth century and the early eighteenth century heavy wartime taxation and generally low levels of prices for agricultural produce weakened their position. Some sank under the weight of accumulated debts, others were ruined by a disastrous harvest or attacks of sheep rot or the cattle plague, yet others decided to sell out and try their fortune in trade or industry—whatever the immediate causes there was a long-term tendency for land to pass from the hands of small owners into the expanding estates of the great owners, and for small properties to become the country residences of the urban professional classes, merchants, tradesmen and manufacturers. 'It is observable', remarked Daniel Defoe when he toured Essex in the early years of the century, 'that in this part of the country, there are several very considerable estates purchas'd, and now enjoy'd by citizens of London, merchants and tradesmen. . . . I mention this, to observe how the present encrease of wealth in the city of London, spreads itself into the country, and plants families and fortunes, who in another age will equal the families of the antient gentry, who perhaps were bought out.'[1]

Throughout our period large estates tended to grow at the expense

[1] D. Defoe, *A Tour through England and Wales* (Everyman ed. 1928) I p. 15.

of small ones. This was inevitable when there were always some small owners whose misfortune, extravagance or ineptitude brought them to sell land; and when there were large owners whose wider sources of income, superior wealth and greater powers of borrowing enabled them to buy more land, and whose strict family settlements meant that only rarely did they sell it again. The large owners were always in the market for the properties of small freeholders and the country gentlemen because new acquisitions would add to their local influence and control of elections, and might help to consolidate their existing property and perhaps make possible a profitable enclosure. The transfer of land was more marked, however, in the periods of low prices and farming difficulties in the first half of the eighteenth century, and again in the years following the end of the Napoleonic wars.

The growth of large estates in the hands of a class of landlords eager to exploit their revenue-yielding potentialities was of course a factor in enclosure, engrossing and improved cultivation. When land was accumulated in one man's hands (even hands tied to some considerable extent by the restrictions on selling, borrowing and land-use imposed in family settlements), it was obviously easier to arrange the substitution of enclosed, compact and consolidated farms for the small, intermingled and inconvenient holdings of the open fields. There were important limitations to the landlords' ability to achieve this transformation, which we shall note when we come to consider the enclosure movement, but nevertheless the encouragement which the growth of larger units of land-ownership gave to agricultural progress should not be overlooked.

Another consequence of the growth of large estates was the increase in tenant farming, and in the course of time there was some tendency for small tenancies to be amalgamated in the hands of large farmers. There is a good deal of evidence in contemporary writings that technical innovation was most notable on the large consolidated properties of the country gentlemen and better freeholders, who had the resources and the spur of self-interest to undertake experiments. But the large tenant-farmers of the landlords were also important as progressive farmers, although we hear less about them. Indeed, as farmers they enjoyed some advantages over the owner-occupiers. The tenants of reputable landlords had little fear of being turned out, even when they had no lease; and the bargaining power of farmers capable of managing large farms enabled them to obtain their land at a moderate rent and obliged the landlords to provide suitable buildings, fences, roads, protection against flooding, and perhaps a part of the expenses incurred in marling or manuring the soil. These large tenant-farmers could devote their own capital entirely to stocking the farm and buying good beasts and equipment; and if the times proved unfavourable and they made losses, the landlords could often be induced to abate their rents and help them through a lean period.

The English system of landlord and tenant, therefore, was one of partnership in which the costs and risks of farming were shared. The landlords provided the basic necessities for good farming, ideally an enclosed, convenient farm, with the buildings well maintained and the land in good heart, and the farmer provided the stock and working capital, and the vital element of skill and enterprise. And in exceptional circumstances the landlord might keep his tenants up to the mark by a carefully detailed and strictly-enforced lease, or might bind the tenants to follow the advice of his steward in the management of the land. Of course many estates fell short of this ideal, the farms too small, run-down and inconvenient, the farmers backward, ignorant or lazy, their landlords incompetent or indifferent. But at its best the English landlord-tenant system was reasonably efficient and flexible—far more so than the conservative peasant farming of the continent—and it provided the essential framework for the great leap forward of agriculture in the eighteenth and nineteenth centuries.

There exists a deep-rooted misconception that before the age of rapid enclosure and 'commercial farming' of the later eighteenth century, the great majority of occupiers farmed mainly for 'subsistence', having little or nothing to do with markets and prices. A moment's reflection, however, must convince that such a view of English farming as it existed before 1760 is mistaken. There was indeed a numerous class of small cultivators who depended on their few acres for a living or as a supplement to their wages, and who had only a small surplus to exchange for necessities which they could not produce themselves. But a very few acres would keep a family in bread corn, and a very few more would support a cow, some pigs and poultry. The farmers proper, i.e. the occupiers of more than a dozen or so acres, must have had considerable surpluses even on quite small farms, and indeed could only have paid their rent and bought their new stock, seeds and implements from the sale of their produce. And furthermore, the size of the non-agricultural population, the existence of flourishing markets in every town of any size, and our knowledge of the agricultural specialization which existed on all sides show that farming was already very well established on a strongly commercial basis at the opening of our period.

When Daniel Defoe wrote his celebrated *Tour of England and Wales* in the 1720s the specialization of the various parts of the country in particular crops, in rearing or in fattening, in dairying, fruit and hops, was already very marked. The bread grains, wheat, barley, rye and oats were grown very widely, it is clear, but with considerable regional concentration. London, the great market, drew its wheat and malt from all the neighbouring counties and also from East Anglia, but Defoe noted Bedfordshire and Hertfordshire as a principal granary, and what was not carried to London by land from Bedford, Hitchin and Hertford, was shipped by river to King's Lynn and then by sea to Holland. Except for London, the greatest corn market was at Farnham in Surrey, where Defoe was told so improbably large a number as 1,100 teams of horses, all drawings loads of wheat, might be seen on market day. From Farnham the corn was ground at Guildford and then sent by barge to London. The industrial areas of the Midlands and Yorkshire drew their grain from their immediate areas and the eastern counties, while the cloth, coal-mining and iron-working areas of Gloucestershire and the seaport of Bristol relied on the corn of Herefordshire and Monmouthshire, some of which was exported to Portugal.

In the Midlands, East Anglia, and near London, large districts were given over to fattening the cattle brought down from Wales and Scotland for the London market. St Faiths, a little village north of Norwich, was the most important mart for the Scots cattle brought to Norfolk for fattening. 'These Scots runts, so they call them', remarked Defoe, 'coming out of the cold and barren mountains of the Highlands in Scotland, feed so eagerly on the rich pastures in these marshes, that they thrive in an unusual manner, and grow monstrously fat; and the beef is so delicious for taste, that the inhabitants prefer 'em to the English cattle, which are much larger and fairer to look at, and they may very well do so: Some have told me, and I believe with good judgment, that there are above 40,000 of these Scots cattle fed in this country every year, and most of them in the said marshes between Norwich, Beccles and Yarmouth.'[1] In the vale of Aylesbury 'all the gentlemen hereabouts are graziers, tho' all the graziers are not gentlemen', stated Defoe, and he was shown one particular pasture field said to be let, again rather improbably, for as much as £1,400 a year. And in the Essex marshes near Tilbury were kept large numbers of Lincolnshire and Leicestershire sheep, bought fat in the autumn and kept back until prices rose in London near Christmas time.

The greatest fair for sheep was held at Weyhill near Andover, where

[1] Defoe, *op. cit.* I p. 65.

said Defoe 'the open down country begins'. 'I confess, though I once saw the fair', he went on, 'yet I could make no estimate of the number brought thither for sale; but asking the opinion of a grasier, who had used to buy sheep there, he boldly answered, There were many hundred thousands. This being too general, I press'd him farther; at length he said, He believed there were five hundred thousand sheep sold there in one fair. Now, tho' this might, I believe, be too many, yet 'tis sufficient to note, that there are a prodigious quantity of sheep sold here; nor can it be otherwise, if it be considered, that the sheep sold here, are not for immediate killing, but are generally ewes for store sheep for the farmers, and they send for them from all the following counties, Berks, Oxford, Bucks, Bedford, Hertford, Middlesex, Kent, Surrey and Sussex: The custom of these farmers, is, to send one farmer in behalf of (perhaps) twenty, and so the sheep come up together, and they part them when they come home.'[1]

From Gloucestershire and Wiltshire great quantities of cheese, bacon and malt were brought down the Thames in barges for the delectation of the London housewives and their families. Stilton cheese was also carried down from Huntingdonshire and was called the English Parmesan, brought to the table said Defoe, 'with the mites, or maggots round it, so thick, that they bring a spoon with them for you to eat the mites with, as you do the cheese.' Another delicacy was Cheddar cheese and Defoe described its making for his readers. The village of Cheddar, he said, had a large green or common

in which the whole herd of the cows, belonging to the town, do feed; the ground is exceeding rich, and as the whole village are cowkeepers, they take care to keep up the goodness of the soil, by agreeing to lay on large quantities of dung for manuring, and inriching the land.

The milk of all the town cows is brought together every day into a common room, where the persons appointed, or trusted, for the management, measure every man's quantity, and set it down in a book; when the quantities are adjusted, the milk is all put together, and every meal's milk makes one cheese, and no more; so that the cheese is bigger, or less, as the cows yield more, or less milk. By this method, the goodness of the cheese is preserved, and, without all dispute, it is the best cheese that England affords, if not, that the whole world affords.

As the cheeses are, by this means, very large, for they often weigh a hundred weight, sometimes much more, so the poorer inhabitants, who

[1] *Ibid.* I p. 289.

have but few cows, are obliged to stay the longer for the return of their milk; for no man has such return, 'till his share comes to a whole cheese, and then he has it; and if the quantity of his milk deliver'd in, comes to above a cheese, the over-plus rests in account to his credit, 'till another cheese comes to his share; and thus every man has equal justice, and though he should have but one cow, he shall, in time, have one whole cheese. This cheese is often sold for six pence to eight pence per pound, when the Cheshire cheese is sold but for two pence to two pence halfpenny.[1]

Norfolk and Suffolk were famous for their turkeys and geese, which every August began the long journey to the capital, proceeding in carts or on foot in droves a thousand or more strong, and feeding on the harvest stubbles as they marched. 'They have counted 300 droves of turkeys ... pass in one season over Stratford-Bridge on the River Stour, which parts Suffolk from Essex, about six miles from Colchester on the road from Ipswich to London', said Defoe.

These droves, as they say, generally contain from three hundred to a thousand each drove; so that one may suppose them to contain 500 one with another, which is 150,000 in all; and yet this is one of the least passages, the numbers which travel by New Market-Heath, and the open country and the forest, and also the numbers that come by Sudbury and Clare, being many more.

For the further supplies of the markets of London with poultry, of which these countries particularly abound: They have within these few years found it practicable to make the geese travel on foot too, as well as the turkeys; and a prodigious number are brought up to London in droves from the farthest parts of Norfolk. . . . They begin to drive them generally in August, by which time the harvest is almost over, and the geese may feed in the stubbles as they go. Thus they hold on to the end of October, when the roads begin to be too stiff and deep for their broad feet and short legs to march in.

Besides these methods of driving these creatures on foot, they have of late also invented a new method of carriage, being carts form'd on purpose, with four stories or stages, to put the creatures in one above another, by which invention one cart will carry a very great number; and for the smoother going they drive with two horses a-breast, like a coach ... changing horses they travel night and day; so that they bring the fowls 70, 80, or 100 miles in two days and one night. . . . [2]

[1] Defoe, *op. cit.* I pp. 277–8.
[2] *Ibid.* I pp. 59–60.

Milk, owing to its nature, could not be transported more than ten miles so that dairies were situated near or actually within the large towns, but London's butter came from as far afield as Suffolk, Yorkshire and Newcastle, although Epping butter was highly popular. The milk for the city was sold from town dairies or direct from the udders of cows driven through the streets. Hygiene was not an advantage of this system, as Smollett's account makes clear. The milk was

> carried through the streets in open pails, exposed to foul rinsings discharged from doors and windows, spittle, snot and tobacco quids, from foot passengers; overflowings from muck carts, spatterings from coach wheels, dirt and trash chucked into it by roguish boys for the joke's sake, the spewings of infants, who have slobbered in the tin-measure, which is thrown back in that condition among the milk, for the benefit of the next customer; and, finally, the vermin that drops from the rags of the nasty drab that vends this previous mixture, under the respectable denomination of milkmaid.[1]

The growth in the eighteenth century of the textile, coal, iron, hardware, and pottery industries, and the concentration of these industries in Lancashire and the West Riding, the Black Country, the north Midlands and the north-east coast, was bound to affect the development of northern agriculture. Trade of course grew with the industries, and Liverpool, Manchester, Hull, Leeds and Newcastle rose greatly in importance and population. Indeed, in a similar way to the great consuming, commercial and manufacturing centre of London, the northern manufacturing towns and villages, the seaports, river-ports and centres of canal transport, extended a pervasive and far-reaching influence over local agricultural specialization.

Northern agriculture was necessarily influenced by climate and relief, and the Pennines, Cheviots and Westmorland uplands could only be made to support the sheep whose wool helped to meet the demand of the West Riding and lesser centres of cloth production, and whose mutton provided staple fare for the Durham and Northumberland coal miners. But wheat, barley and oats were widely grown in the East Riding, in parts of Lancashire and in the north Midlands, and as the northern population grew so the farmers of the grasslands found it increasingly profitable to buy Scots and Irish cattle to fatten for the local markets. Large areas of Westmorland, Cumberland, Lancashire and the Craven district of Yorkshire—especially well situated for supplying industrial Lancashire and Yorkshire to the west and south and the Yorkshire ports

[1] T. Smollett, *The Expedition of Humphry Clinker* (1771).

25

to the east—were given over to this trade. The dairies of Cheshire, Warwickshire and Lancashire, while concerned in large part with the London and Irish markets, were more and more influenced by the demand from the swelling industrial population on their doorsteps, and as potatoes became an increasingly popular supplement to the northern oatcakes and bannocks, so potato-growing developed on a large scale in Lancashire and Cheshire, and quantities were shipped from Liverpool to Dublin. Other vegetables, too, were cultivated, and the dairy farmers of Cheshire and the neighbouring areas sent to market great quantities of pork and bacon.

The demand of the domestic textile workers, coal miners and others for beasts used as pack-animals, and for hauling wagons and working horse-gins, was important for the great northern trade in horses and oxen. 'Every clothier', remarked Defoe, 'must keep a horse, perhaps two, to fetch and carry for the use of his manufacture (*viz.*) to fetch home his wooll and his provisions from the market, to carry his yarn to the spinners, his manufacture to the fulling mill, and, when finished, to the market to be sold, and the like.'[1] The North Riding in particular was famous for its horses, bred also as hunters for sportsmen and chargers for the army. And even in Defoe's time the influence of the manufacturing population of the West Riding was already plainly visible on the agriculture of the neighbouring counties:

> the West Riding is thus taken up, and the lands occupied by the manufacture; the consequence is plain, their corn comes up in great quantities out of Lincoln, Nottingham and the East Riding, their black cattle and horses from the North Riding, their sheep and mutton from the adjacent counties every way, their butter from the East and North Riding, their cheese out of Cheshire and Warwickshire, more black cattle also from Lancashire. And here the breeders and feeders, the farmers and country people find money flowing in plenty from the manufacturers and commerce; so that at Halifax, Leeds, and other great manufacturing towns so often mentioned, and adjacent to these, for the two months of September and October a prodigious quantity of black cattle is sold.[2]

It is evident then that the demand for food and raw materials of London, the growing seaports like Liverpool and Bristol, and the manufacturing areas of Yorkshire, Lancashire, East Anglia and the west country, encouraged a great degree of specialization and influenced the nature of

[1] Defoe, *op. cit.* II p. 195.
[2] *Ibid.* II p. 199.

farming at great distances from the ultimate markets. As a result there was an enormous amount of transporting of agricultural produce, for which the rivers and coastal shipping were indispensable. The East Riding's surplus corn was carried on the local river system to Hull and there shipped to London, Holland, Hamburg, France and Spain, while the barley and malt of the north-east Midlands was carried up the Trent to the brewers of Burton, and into Derbyshire, Cheshire and Lancashire. The Black Country's corn was supplied from Staffordshire, Shropshire. Herefordshire, Buckinghamshire and the Vale of Evesham. Wool carried south-westwards from Lincolnshire, Leicestershire and Northampton-shire was sold at Cirencester and other west country clothing towns in order to supplement the local supply and that coming eastwards through Bristol and Gloucester from Hereford and the Welsh marches. Great quantities of hops were sold at the famous Stourbridge fair, held in a large corn field near Casterton between Cambridge and Newmarket. The hops were transported there by water from the hop grounds of Kent and Surrey via the port of King's Lynn and the rivers Ouse and Cam, and the hops, when sold, were carried away into all the neighbouring counties or shipped back to King's Lynn by the same route to be sent by sea to Hull and Newcastle for the northern counties, and even to Scotland.[1]

Cider from Devonshire, and even from Herefordshire,[2] went by sea right round Land's End to London, and the cheesemongers of the capital owned sixteen vessels for bringing cheese round by sea from Cheshire. The 'Cheshire' cheese, made also in Shropshire, Staffordshire and Lancashire, was sent down the Severn to Bristol, and sailed also down the Trent to Gainsborough, York and Hull, while quantities were shipped from the Mersey to Ireland. According to Defoe, 4,000 tons of Cheshire cheese were carried down the Trent to Hull, some intended for Scotland and the northern English counties, the remainder for London by the shorter east coast alternative to the western sea route, 'a terrible long, and sometimes dangerous voyage, being thro' the Irish Channel, round all Wales, cross the Bristol Channel, round the Land's End of Cornwall, and up the English Channel to the mouth of the Thames, and so up to London. . . . Again, the Gloucestershire men carry all [their cheese] by land-carriage to Lechlade and Cricklade on the Thames, and so carry it down the river to London. But the Warwickshire men have no water-carriage at all, or at least not, 'till they have carry'd it a long way by land

[1] *Ibid.* I pp. 82-3
[2] E. L. Jones, 'Agricultural Conditions and Changes in Herefordshire, 1660–1815 *Trans. Woolhope Club* 37 (1961) p. 33 n. 5.

to Oxford; but as their quantity is exceeding great, and they supply not only the City of London, but also the counties of Essex, Suffolk, Norfolk, Cambridge, Huntingdon, Hertford, Bedford and Northampton . . . in all which cases land-carriage being long, and the ways bad, makes it very dear to the poor, who are the consumers.'[1]

The improvement of the main roads by turnpikes was a novel and remarkable feature of Defoe's time and attracted his especial attention. Better-constructed roads with hard surfaces were much needed in the areas of soft deep clays of the Midlands and home counties, where the soft going had long obstructed transport and in the winter was 'perfectly frightful to travellers.'

> The reason of my taking notice of this badness of the roads, through all the midland counties, is this; that as these are counties which drive a very great trade with the City of London, and with one another, perhaps the greatest of any counties in England; and that, by consequence, the carriage is exceeding great, and also that all the land carriage of the northern counties necessarily goes through these counties, so the roads had been plow'd so deep, and materials have been in some places so difficult to be had for repair of the roads, that all the surveyors rates have been able to do nothing; nay, the very whole country has not been able tc repair them; that is to say, it was a burthen too great for the poor farmers; for in England, it is the tenant, not the landlord, that pays the surveyors of the highways.
>
> This necessarily brought the country to bring these things before the Parliament; and the consequence has been that turnpikes or toll-bars have been set up on the several great roads of England, beginning at London, and proceeding thro' almost all those dirty deep roads, in the midland counties especially; at which turn-pikes all carriages, droves of cattle, and travellers on horseback, are oblig'd to pay an easy toll; that is to say, a horse a penny, a coach three pence, a cart four pence, at some six pence to eight pence, a waggon six pence, in some a shilling, and the like; cattle pay by the score, or by the head, in some places more, in some less; but in no place is it thought a burthen that ever I met with, the benefit of a good road abundantly making amends for that little charge the travellers are put to at the turn-pikes.[2]

Defoe went on to detail the advantages of the road improvements:

> as for trade, it will be encourag'd by it every way; for carriage of all kind of

[1] Defoe, *op. cit.* II pp. 131, 141.
[2] *Ibid.* II pp. 118–19.

heavy goods will be much easier, the waggoners will either perform in less time, or draw heavier loads, or the same load with fewer horses; the pack-horses will carry heavier burthens, or travel farther in a day, and so perform their journey in less time; all which will tend to lessen the rate of carriage, and so bring goods cheaper to market.

The fat cattle will drive lighter, and come in market with less toil, and consequently both go farther in one day, and not waste their flesh, and heat and spoil themselves, in wallowing thro' the mud and sloughs, as is now the case.

The sheep will be able to travel in the winter, and the city not be oblig'd to give great prizes to the butchers for mutton, because it cannot be brought up out of Leicestershire and Lincolnshire, the sheep not being able to travel: the graziers and breeders will not be oblig'd to sell their stocks of weathers cheap in October to the farmers within 20 miles of London, because after that they cannot bring them up; but the ways being always light and sound, the grasiers will keep their stocks themselves, and bring them up to market, as they see cause, as well in winter as in summer . . . were the roads made good and passable, the city would be serv'd with mutton almost as cheap in the winter as in the summer. . . .[1]

Among the improved roads mentioned by Defoe was that from London to Ipswich and Harwich, formerly 'the most worn with waggons, carts, and carriages, and with infinite droves of black cattle, hogs, and sheep, of any road'; another passed northwards from London to Bedfordshire through Baldock Lane, once 'famous for being so unpassable, that the coaches and travellers were oblig'd to break out of the way even by force, which the people of the country not able to prevent, at length placed gates, and laid their lands open, setting men at the gates to take a voluntary toll, which travellers always chose to pay, rather than plunge into sloughs and holes, which no horse could wade through.'[2]

Cattle and sheep, as Defoe makes clear, travelled on their own legs, and often over vast distances. The fattening areas of the northern counties, East Anglia, the Midlands and home counties, and ultimately the London markets, were the main goals of the drovers. Sheep bought at Burford fair were fattened in every county between Oxfordshire and Kent, while Scots and Welsh sheep were driven hundreds of miles into the home counties to become London's mutton. Scots cattle travelled south on either side of the Pennines and were fattened in the northern counties,

[1] *Ibid.* II pp. 127, 130.
[2] *Ibid.* II pp. 121–3.

Lincolnshire, Leicestershire and East Anglia—most of the fat cattle sent out of Norfolk were Scots or Irish.

The influx of Scots cattle only became very large after the Act of Union, but the great trade in Welsh cattle goes back to the Middle Ages. Anglesey cattle swam the Menai Strait as an adventurous prelude to their pilgrimage to Barnet fair and the scene was graphically described by a late eighteenth-century writer:

> They are urged in a body by loud shoutings and blows into the water and as they swim well and fast, usually make their way to the opposite shore. The whole herd proceeds pretty regularly until it arrives within about 150 yards of the landing-place, when, meeting with a very rapid current formed by the tide eddying and rushing with great violence between the rocks that encroach far into the Channel, the herd is thrown into the utmost confusion. Some of the boldest and strongest push directly across and presently reach the land; the more timorous immediately turn round and endeavour to reach the place from which they set off; but the greatest part, borne down by the force of the stream, are carried towards Beaumaris Bay and frequently float to a great distance before they are able to reach the Caernarvonshire shore.[1]

However, some of the cattle from Pembrokeshire, Carmarthen and Brecknock travelled in more gentlemanly fashion by ship across the Severn, or were sailed from Tenby to the Somerset ports. The chief route from south and central Wales was the land one through Hereford-shire, Ross, Ledbury and Tewkesbury to the south midland grasslands, and even as far as Essex and Kent. The favourite route from North Wales was through Castle Bromwich and Warwickshire in the direction of Barnet fair, the Essex grazings, and the London butchers. The drovers could cover up to 15 or 20 miles a day and cared little about distance, preferring a circuitous route through the lanes and byways rather than pay the tolls of the high roads, but the expense of shoeing the cattle in preparation for their great march could not be avoided. We are told, not surprisingly, that the local farmers hastily moved their own cattle to a safe distance when it was rumoured that a Welsh drove was on the way.[2]

The small black Scots cattle were driven many hundreds of miles from their breeding grounds in northern and north-western Scotland. The drovers went up into the Highlands to buy their cattle in April and May,

---

[1] Aikin's *Journal of a Tour from North Wales in 1787*, quoted by A. R. B. Haldane, *The Drove Roads of Scotland* (1952) pp. 183–4.

[2] C. Skeel, 'The Cattle Trade between Wales and England from the Fifteenth to the Nineteenth Centuries', *Trans. Roy. Hist. Soc.* IX (1926) pp. 142–9.

and from then until autumn the droves of 100 to 300 beasts wended their way over moor and mountain and through the glens, covering some 10 to 12 miles a day in making for the Lowland 'trysts' or fairs held at Crieff and Falkirk. From these fairs the cattle not sold locally were taken south across the border in droves often 1,000 strong.

The Hebrides, and particularly Skye, were the home of the earliest supplies of cattle for the fairs of southern Scotland and England, and Kyle Rhea, the narrowest channel between Skye and the mainland, was always the principal crossing point from the island. In similar fashion to the Welsh beasts crossing over from Anglesey, the Skye cattle were forced to swim this dangerous passage with its strong current and inshore eddies. Nevertheless, few beasts were lost, the drovers taking care to keep the beasts' tongues free by means of ropes placed round their lower jaws. 'The reason given for leaving the tongue loose is that the animal may be able to keep the salt water from going down its throat in such a quantity as to fill all the cavities in the body which would prevent the action of the lungs; for every beast is found dead and said to be drowned at the landing place to which this mark of attention has not been paid. Whenever the noose is put under the jaw all the beasts destined to be ferried together are led by the ferryman into the water until they are afloat, which puts an end to their resistance. Then every cow is tied to the tail of the cow before until a string of 6 or 8 be joined. A man in the stern of the boat holds the rope of the foremost cow. The rowers then ply their oars immediately.'[1]

Like their beasts, the drovers had to be strong and hardy, and they were accustomed to settling down to sleep by the cattle on the coldest of nights with only a plaid for a covering, their ears attuned to waken at the slightest sound of straying. Sometimes in the course of the long southwards journeys the cattle were bled for the essential ingredient of the drovers' black puddings, which with oatmeal and a ram's horn of whisky, provided their staple fare.[2] The droving trade was subject to severe and unpredictable losses, and called for a willingness to undertake financial as well as physical risks. Most drovers had little capital and relied on credit to finance their purchases of Highland cattle, and they were often involved in short-term debts of more than £10,000 in a single season. Prices could fall disastrously while the beasts were on the road, leaving the drover with insufficient money when he sold his cattle to cover his expenses and meet the bills or promissory notes that he had given the breeders in the Highlands. The cattle plague which raged in England in the middle years of the

[1] Haldane, *op. cit.* pp. 28–9, 35, 69–77.
[2] *Ibid.* pp. 27, 42.

eighteenth century and occasionally thereafter was another grave hazard for the drovers.[1] Despite their large-scale credit transactions the Scottish drovers, unlike their Welsh counterparts, do not seem, however, to have become carriers of money, and ultimately, bankers.

At the time of the Act of Union, 1707, at least 30,000 cattle were being driven across the border every year, and as the demands of London, the growing industrial towns, and the armed forces increased, so the droving trade grew and prospered. By the middle eighteenth century as many as 80,000 cattle and 150,000 sheep threaded their way through the Cheviots, entering England through Eskdale, Liddesdale and Carlisle or across the shallow Solway Firth. Particularly important for the droving trade became the rising purchases of the graziers of Northumberland and north-west Yorkshire who supplied the expanding population of Tyneside, Hull and the Yorkshire coast, 'while the grass lands of Cumberland, Westmorland and the Craven district of Yorkshire fed the cloth workers of the West Riding and the mill workers of Lancashire.'[2] Others pressed on farther south: 'From Northallerton their road lay by Boroughbridge and Doncaster and on by Gainsborough and Newark to Grantham and Peterborough, following roughly the route of the Great North Road which a traveller of the middle of the eighteenth century described as having wide stretches of turf on either side perpetually roughened by the passage of great droves of cattle. Some, turning east about Grantham, went by Spalding and Wisbech into Norfolk, and a record of 1750 reports 20,000 Scots cattle as passing along the Wisbech road. The routes of the drovers through Yorkshire and Lincolnshire, as in other parts of the north and Midlands of England, are marked by the names of wayside inns, some of which still survive. The Drover's Inn at Boroughbridge and Wetherby, the Drover's Call between Gainsborough and Lincoln and the Highland Laddie at Nottingham and St Faith's near Norwich, recall the droving traffic to East Anglia and the home counties, as the Highland Laddie at Gretna and the Drover's Rest at Kirkandrews in Cumberland recall that bound for Cumberland and the West Riding.'[3] Whatever route they took the drovers tried as far as possible to keep to the higher ground and moorlands, avoiding hard roads which injured the cattle's feet and the highly cultivated and enclosed areas where the passage was narrow, wayside grazing scarce and nightly 'stances' for the beasts to be had only at a price.

[1] Haldane *op. cit.* pp. 47–9, 67.
[2] *Ibid.* pp. 173, 178.
[3] *Ibid.* p. 180.

The ultimate destination for most of the cattle which travelled so far as the Midlands and East Anglia was of course London, and it is apparent from Defoe's account that London, with its half a million consumers and much the largest centre of population in the country, was the great centre of agricultural trade. Probably something over 300,000 quarters of corn were required in London every year, and over 75,000 cattle and 500,000 sheep were annually sold at Smithfield. London was the great determinant of agricultural specialization, of trade routes, and to a considerable extent of the ruling level of prices. But in addition the other centres of large consumption also created markets and specialized demands, as we have seen, and the costs of transporting produce over considerable distances, as well as variations in weather conditions, gave rise to some local monopolies and local variations in prices. The general picture is one of an agriculture geared to the needs of a variety of markets, and particularly to that of London. It is a picture, as Defoe composed it, of bustling activity, prosperity, experiment and mobility. Of course it should be remembered that Defoe had a special interest in trade and was concerned to stress its importance and extent. However, as Professor Ashton has commented, 'Defoe had an eye for whatever was striking or unusual, and sometimes he ran to hyperbole; but it is impossible to ignore the picture he presents. It is certainly not one of a community of supine peasants.'[1]

*References and suggestions for further reading:*

| | |
|---|---|
| T. S. Ashton | *An Economic History of England: the Eighteenth Century* (1955), Ch. II. |
| D. Defoe | *A Tour through England and Wales* (Everyman ed. 1928). |
| G. E. Fussell & C. Goodman | 'Traffic in Farm Produce in Eighteenth-Century England' *Agricultural History* XII, 4 (1938). |
| G. E. Fussell & C. Goodman | 'Eighteenth-Century Traffic in Livestock' *Economic History* III (1936). |
| H. J. Habakkuk | 'English Landownership, 1680–1740' *Economic History Review* 2nd ser. X (1940). |
| A. R. B. Haldane | *The Drove Roads of Scotland* (1952). |
| G. E. Mingay | *English Landed Society in the Eighteenth Century* (1963), Ch. I–IV. |

[1] T. S. Ashton, *An Economic History of England: the Eighteenth Century* (1955) p. 33.

# 2 The Eighteenth Century and Improvement

The eighteenth century, of course, has always been known as a period of remarkable expansion and improvement in agriculture. However, we have no very accurate figures for the average yields of crops or the total numbers and weight of beasts sent to market, and the evidence for a substantial increase in output is essentially of a circumstantial kind. We can draw on the writings of contemporary experts like Young, Kent and Marshall to show that innovation was in the air and that many technical improvements were being adopted by the better farmers; and the fact of the enclosure movement shows that it was increasingly profitable to use land more economically and for the type of produce to which it was naturally best suited. By relating our knowledge of the growth of population and industry with the movements of agricultural prices and changes in overseas trade in agricultural produce, it is possible to hazard a guess at the magnitude of the overall increase in output: in view of the rise in prices and the shift to the position of a country importing significant quantities of grain, cattle and dairy produce, it would appear that output in England and Wales must have increased at something rather less than the rate of growth of population—an increase of the order of 40 or 50 per cent would seem to meet the change in the economic situation.

Our very rough estimate fits in well with the figures that Phyllis Deane and A. W. Cole, using much more complex methods of calculation, have arrived at. They estimate that output of agricultural produce increased by 43 per cent over the eighteenth century. Of the main components of output they calculate that the overall increase in grain production of England and Wales (estimated at 43 per cent) was achieved by a rise in yields per acre of rather more than 10 per cent and an increase in the sown acreage of perhaps 25 per cent. Animal husbandry, they think, grew unevenly: the increase in wool and mutton was much greater than in grain production—the wool clip nearly doubled—while beef and dairy production probably only kept pace with the growth of population.[1]

Owing to the doubtful factors in the methods of calculation we cannot put too much reliance on these estimates, but it is significant that the rise in grain output is believed to be due more to the increase in the sown acreage than to an improvement in yields. In the first of these factors enclosure was the vital instrument. Where enclosure led to the abolition of fallows, the effective acreage of newly-enclosed parishes was increased by up to a third or a half; it should be borne in mind, nevertheless, that quite often fallowing continued after enclosure, and that even in open-field villages not all the land—sometimes as little as half of it—lay in open fields and was subject to fallowing. However, many enclosures were not concerned with open fields alone but were directed partly or entirely towards the proper cultivation of overstocked or neglected commons, or the bringing into production for the first time of waste land from moors, marshes and forests. It has been estimated that between 1760 and 1799 enclosures brought at least 2,000,000 acres, and perhaps over 3,000,000 acres, of waste land into cultivation.[2] This was mainly in the northern counties, Yorkshire, Northumberland, Cumberland and Westmorland, but also in Lincolnshire, Derbyshire and Wales, and in these areas there still remained at the end of the century many thousands of acres of waste suitable for some form of agricultural use. There was also much enclosure of waste, drainage of marshes and deforestation in the earlier years of the century, but in this period the spread of cultivation was perhaps most marked in the light soil areas and chalk and limestone uplands of the southern half of the country, which by the adoption of convertible husbandry could be made more profitable than serving merely as sheep runs.

Of probably rather less importance in the rise of output in the eighteenth century was the more efficient and intensive cultivation which followed the enclosure of open fields and commons and the achievement of a better balance between arable land and pasture, the creation of more convenient and larger farms, and the use of better rotations with roots and legumes, improved breeds of sheep, cattle and horses, and more efficient implements. There is also no reliable way of estimating accurately this increase

[1] Deane & Cole, *op. cit.* pp. 68–70, 78–9.
[2] G. R. Porter, *The Progress of the Nation* (2nd ed. 1851) p. 157.

in output due to improved farming methods. We know that the yield per acre of wheat varied very greatly according to the type of soil, but what the national average yield was, and whether it increased by only as much as 10 per cent, from 20 to 22 bushels per acre as has been suggested,[1] fall into the realm of speculation. Similarly, little can be said with certainty about the advance in livestock production. The figures for sales of cattle and sheep at Smithfield market show a substantial rise of about 40 per cent between 1732 and 1794, but this no doubt exaggerates the increase in the supply of cattle, beef being more in demand than mutton in the London market. Most of the increase in mutton and beef supplies came from an increase in the average carcass weight owing to the development of improved breeds with more meat and less bone and offal, and larger and more varied supplies of fodder, but it is not now accepted that this change was as revolutionary as once thought. An important factor in supply was the large and increasing numbers of cattle and sheep driven across the Scottish border for fattening in England. Again there are no reliable national figures, and the appearance of a marked change in the size of cattle brought to market which so impressed some contemporaries, may have resulted in part from the abandonment of the former practice of sending only the smaller beasts to market and keeping the larger cattle on the farm for draught purposes. Certainly the advances in breeding associated chiefly with the names of Bakewell and the Collings brothers were aimed at producing heavier and more rapidly fattened animals, but it is doubtful whether they could have made a very widespread impression on the general standards of livestock much before the end of the century.[2] From the evidence of supplies of wool, and assuming that there was on balance a growth of flocks associated with the spread of convertible husbandry on light soils in the early part of the century, it seems likely that mutton production did perhaps increase rather more than did beef and dairy produce, as is argued by Deane and Cole.[3]

The difficulties of assessing the extent of improvement in the eighteenth century are intensified by the great differences in techniques and progress which marked off one part of the country from another. According to Marshall, the west country and northern methods had certain peculiarities, and were generally cruder and less efficient than those in use in the

[1] G. E. Fussell, 'Population and Wheat Production in the Eighteenth Century' *History Teachers Miscellany* VII (1929) p. 88.

[2] See G. E. Fussell, 'The Size of English Cattle in the Eighteenth Century' *Agricultural History* III (1929) pp. 160–81, and the discussion in Deane and Cole, *op. cit.* pp. 68–74.

[3] *Op. cit.* pp. 68–73.

Midlands and eastern counties. In Devonshire, for example, the practice of 'winnowing with the natural wind' attracted Marshall's astonished attention:

> Farmers of every class (some few excepted) carry their corn into the field, on horseback, perhaps a quarter of a mile, from the barn, to the summit of some airy swell; where it is winnowed, *by women*! the mistress of the farm, perhaps, being exposed in the severest weather, to the cutting winds of winter, in this slavish, and truly barbarous employment. The obsolete practice of the Northern extremity of the Island, in which farmers loaded their wives and daughters with dung, to be carried to the fields on their backs, was but a little more uncivilized. The machine fan, however, is at length, making its way into the Western extremity.

The Devonshire farmers, Marshall found, tended to be ignorant and conservative, and their labourers often 'drunken, idle fellows'. Many of the farmers had risen 'from servants of the lowest class; and having never had an opportunity of looking beyond the limits of the immediate neighbourhood of their birth and servitude, follow implicitly the paths of their masters. Their KNOWLEDGE is of course confined; and the SPIRIT OF IMPROVEMENT deeply buried under an accumulation of custom and prejudice.'[1]

But the differences in farming practices could even be very marked within a single county, according chiefly to relief and the character of the soil, the nature of the farms and the availability of communications with markets. In Cambridgeshire, for instance, the western districts nearest the Great North Road were more rapidly enclosed and more generally advanced than the less accessible eastern parts of the county; in Leicestershire and Nottinghamshire heavy lands suited to pasture were among the earliest enclosed, while poorer and less rewarding soils were left till much later; and while in Bedfordshire there were plenty of progressive and prosperous farmers on the better soils, the districts of poor cold clays were characterized by small farms and backward methods. There can be little doubt that relief, soil, enclosure and communications were the main determinants of precisely where and when improvement occurred in a particular area, but to regard this generalization as the entire solution to a many-sided problem would be to underestimate the complexity of agricultural development.

At any time the changes in markets and prices, reflecting shifts in demand and supply, were the basic elements in the situation. Price

[1] W. Marshall, *The Rural Economy of the West of England* (1796) I pp. 106–7, 184.

changes were of two broad kinds: the short-term, seasonal and year-to-year movements which largely reflected the influence of weather on the quantities brought to market; and the more long-term shifts in the price level of important commodities such as grain, beef, mutton, wool and dairy produce, which were the result of a gradually changing overall balance between the demand for food and raw materials and the output as determined by land-use and methods of farming.

The short-term price movements were extremely uncertain and variable, because weather conditions varied from one district to another and did not affect all farm products in the same way. It was rare for one type of weather to persist for very long or to have a uniform influence on a particular crop. To some extent weather effects would cancel out. For example, a summer too wet for good crops of grain in the badly-drained clayland vales might be just right for the free-draining upland grain areas. A poor harvest for grain might result not only from unsuitable weather at harvest time but also from bad conditions earlier in the growing season, and particularly from excessive rainfall in the preceding sowing period, resulting in a smaller than normal acreage being sown. The flow of beasts to the markets was greatly influenced, of course, by the effects of weather on fodder supplies. A long and severe winter held back the growth of spring grass and might cause rot among the turnips, so making the 'hungry gap' of March and April more than normally difficult to fill, while a prolonged drought in summer meant low yields of hay and roots, and so short reserves of fodder for the next winter. Whenever a fodder shortage threatened breeders and graziers limited their numbers of lean stock and sent fat beasts early to market. Prices would thus be temporarily depressed, but the reduced size of flocks and herds would mean higher prices in subsequent years. The numbers of pigs coming to market were similarly affected by supplies of natural feed such as acorns, beechmast and apples, while sheep were very much subject to liver-rot in warm, damp autumns, and to a lesser extent to losses through cold and snow in severe winters.[1]

Short-term price fluctuations, therefore, being largely the consequence of weather conditions (which also affected, of course, transport and the means of getting crops and beasts to market), were highly unpredictable. They created great uncertainty for farmers and gave an advantage to those farmers whose soil, situation, and personal enterprise allowed more flexibility in their farming plans. Bad seasons were not usually sufficiently

[1] For an excellent discussion of the extremely varied effects of weather on farming see E. L. Jones, *Seasons and Prices* (1964), from which this brief summary is drawn.

serious, however, to cause widespread bankruptcies, particularly as short supplies in the market led to the compensation of higher prices. Grain farmers, indeed, might more frequently complain of a run of good seasons which lowered prices to unprofitable levels.

The long-term price shifts, on the other hand, were more evident and more easily foreseen, and gave rise to more permanent changes in farmers' techniques and use of the land. A long-term tendency for prices to rise encouraged increased production, more rapid intake of waste lands, and more effective use of the existing farm land. It was no accident that the great acceleration of enclosure, and the publicity given to innovations by contemporary writers, which characterize the period after 1760, had as their background a level of prices that was not only markedly higher than in the first half of the century, but which also showed a tendency to rise fairly steadily (see fig. 2, page 110). A rising level of prices indicated increasing pressure on supply and increasing profit in enclosing land and cultivating it more intensively. It was the generally rising level of prices between about 1760 and 1813—rising fairly slowly at first, but much more rapidly with the shortages after 1793—which encouraged and largely brought about in that period the great increase in the cultivated acreage and the more widespread adoption of better methods of husbandry. Higher prices meant greater profits for farmers and a greater demand for land. Landlords were thus enabled to ask higher rents, and the increased value of properly cultivated land spurred on enclosure. Farmers who had the prospect of making large profits were more willing to contemplate the expense of heavy marling and manuring, undertake the risks of experimenting with new rotations, and to consider paying 20 guineas a season in order to hire one of Bakewell's rams.

In the first half of the eighteenth century the generally lower and often unprofitable level of prices did not encourage such rapid development, but nevertheless often acted as a spur to improvement. When prices were low farmers had to be more efficient in order to survive. Enterprising farmers sought means of reducing their costs of production per unit of output, i.e., they tried to make more economical use of their land and labour force. Indeed, the greater use of fodder crops in rotations, of clover and sainfoin, and roots like turnips, which could be used both as feed for stock and for improving soil fertility, was evidence of the ability of progressive farmers to increase their yield per acre, produce fatter beasts, and have greater flexibility in their cropping plans. Clovers and other grasses were already in wide use in the Midlands and south and were penetrating the East Riding of Yorkshire and other northern districts, but the famous Norfolk four-course rotation, in which crops of wheat and barley alternated with the fodder and soil-restoring crops of turnips and clover, was the most striking development, and was already well established in East Anglia, Hertfordshire, Essex and elsewhere in the early eighteenth century. The light sandy soils of parts of these counties were the ideal environment for growing roots and for the barley which was in great demand by the London and Dutch brewers and distillers, markets which of course were conveniently close by land or easily reached by sea.[1]

As corn prices ruled low in the first half of the eighteenth century the advantages of convertible husbandry were increasingly appreciated, and on the lighter soils dry enough for the winter folding of sheep on roots and grasses, there was conversion of grassland, hill pastures and waste to arable. In Hampshire, for example, the expansion of convertible husbandry on the Norfolk pattern represented attempts to increase output in response to the low prices of grain and wool between 1730 and 1750.[2] The farmers who practised this system were not entirely dependent on grain prices, for a large proportion of their land was devoted to feeding stock on roots and clovers, and so they might have more fat-stock, and wool for sale as well as corn.

Nevertheless, we should be careful not to exaggerate the economic advantages which convertible husbandry gave to light soil farmers. It is evident that the low prices of wheat and barley had a severe impact even on the large and efficient Norfolk farmers. Their rents were high and were governed by long leases, and their labour costs could not be easily reduced when prices fell. In the bad years at the beginning of the century Norfolk farmers had great difficulty in paying their rents, and Lord Stanhope complained in 1702: 'my Tenants never paid my rents so ill as this last half year.' An attorney was sent with instructions to force the farmers to pay up, but this remedy failed miserably, the attorney finding 'that all the corn lyes dead upon their hands so that to seize their persons when they have no money among them will do no good.' Sir Robert Walpole, another Norfolk proprietor, experienced the same trouble, his steward reporting that the tenants' 'great complaint is (and not without reason) the want of a price for their corn.' Subsequently in 1705 several of Sir Robert's tenants went bankrupt and others gave up their farms,

---

[1] A. H. John, 'The Course of Agricultural Change, 1660–1760', in *Studies in the Industrial Revolution* (ed. L. S. Pressnell, 1960) pp. 130–1.

[2] John, *op. cit.* pp. 146–8; E. L. Jones, 'Eighteenth-century Changes in Hampshire Chalkland Farming' *Ag. Hist. Rev.* VIII (1960) p. 9.

and it proved necessary to reduce the rents in order to attract new tenants.[1] Again between 1709 and 1712 when corn prices fell the Norfolk farmers were unable to pay their rents, but the most serious difficulty was experienced in the 1730s and 1740s when wheat averaged only about 30s. a quarter, and barley fell to between 10s. and 15s. Over the whole Coke estates (including some outside Norfolk) arrears of rent which had averaged under £2,000 in the later 1720s increased considerably, and between 1734 and 1736 were well over £4,000—a third of the gross rental.[2]

Nevertheless it may well be that between 1730 and 1750 the Norfolk and other light land farmers, with their productive and flexible four-course rotation, were less severely affected by low grain prices than farmers elsewhere, particularly those with open-field farms in the heavy soil areas of the Midlands. The small open-field farmers' means of increasing output were inevitably limited by the rigidities of the open-field system, and farmers on heavy soils faced higher costs of cultivation, lower yields, and very often greater costs of marketing their produce through the notoriously bad roads of clay districts. Even these farmers, however, made some adjustment to the conditions, exchanging lands to make their farms more compact, converting parts of the open fields into pasture closes, and obliging their landlords to help them with repairs and improvements to farm buildings. Moreover, while their methods of farming were relatively inefficient their rents were low, and they employed little labour beyond that of their own families.

The evidence of depression conditions in this period is remarkably widespread and included light and heavy land, enclosed and open-field farms, and pasture as well as arable. Landlords complained of unpaid rents and bankrupt tenants from areas so diverse as Denbighshire, Cheshire and Gloucestershire in the west, Yorkshire, East Anglia and Lincolnshire in the north and east, down through Derbyshire, Nottinghamshire, Rutland, Staffordshire, Buckinghamshire, Middlesex and Essex and as far south as Kent, Sussex and Hampshire.[3] In Staffordshire the Earl of Dartmouth's agent wrote in 1734 that 'I have never known money so scarce amongst y$^r$ Lordship's tenants since I was concern$^d$', while in the next year a defaulting Gloucestershire farmer wrote to Lord Hardwicke's steward: 'it is a very Rong time to put a tennant to any Charge in Law it being such a time that we tennants can hardly make

---

[1] J. H. Plumb, *Sir Robert Walpole* (1956) pp. 16, 106, 186.
[2] We are greatly obliged to Dr R. A. C. Parker for this information.
[3] See Mingay, *English Landed Society in the Eighteenth Century* pp. 54–6.

anything of the Estates to pay Rent soe that I rely Intirely upon his Lordship's Mercy.'[1] Of course the fat-stock specialists and dairymen benefited directly from the good corn harvests and the cheapness of the grain which they needed to feed their beasts. In the first half of the century fattening and dairying were generally more profitable than corn because of the more stable prices for mutton and beef and for butter, cheese, pork and bacon. It should be noted, however, that wool prices tended downwards, and many dairymen and graziers were hit by unusually hard winters and occasional fodder shortages, and also by the ravages of liver-rot among sheep and the cattle plague after 1745. An important factor in the market situation for pasture farmers was the improvement in living standards at this time, possibly a quite marked improvement, and in part the more stable prices of meat and dairy produce reflected a greater *per capita* consumption of these superior foodstuffs. As a consequence there was a tendency for lands unsuited to convertible husbandry to be put down to permanent grass or leys, as in the wealden area of Kent and Surrey and the clays of Leicestershire and the central Midlands— both for dairying and for fattening the larger numbers of lean stock brought into England from Wales and Scotland.[2]

In these conditions of generally low prices before 1760, outbreaks of animal disease, killing winters, and periodical harvest failure, as in 1710 and 1740, it is not surprising that numbers of the inefficient and less enterprising farmers failed. There was a weeding-out process, and particularly in areas of the Midlands the numbers of small farmers declined sharply, their holdings going to swell the acreages of the more efficient and financially sounder occupiers of larger farms. Landlords, as we have remarked already, were always willing to buy up small properties. Such purchases were often made with political rather than agricultural objects in mind, or merely in order to extend the boundaries of an estate. But often the holdings were bought to improve the layout of the existing farms, make them larger, more compact and easier to work, and perhaps with the ultimate aim in view of enclosure of the whole or part of the open fields. Landlords anxious to improve their rentals might take steps to enclose waste land, or might help farmers to improve their output by meeting part of the cost of marling thin soils or of draining boggy ones.

---

[1] William Salt Library, Stafford: D. 1778 V 1210; Gloucester Record Office: D 214 E 18.

[2] John, *op. cit.* p. 149. See also the same writer's valuable article: 'Aspects of English Economic Growth in the First Half of the Eighteenth Century' *Economica* May 1961 pp. 180–3.

The farmers of open fields, as we have noted, often co-operated in arranging among themselves small enclosures, exchanges of lands, agreements to sow the fallow field with sainfoin, clover or turnips, or regulations to stint an overstocked common. These gradual piecemeal improvements, and the engrossing of farms, certainly went on throughout the century, but may well have been more common in the first half of it.[1]

The devious path of agricultural improvement was influenced also by political and legal changes. The Act of Union in 1707, as we have noticed, increased the flow of Scots cattle to the English fattening grounds; the import of Irish pastoral products, on the other hand—cattle, sheep, pork and bacon, butter and cheese—was prohibited by legislation of Charles II and William III, and the prohibition held good until the third quarter of the eighteenth century when increasing pressure of population on food resources pushed up prices, created concern and discontent, and led to relaxations in Irish imports; and the Corn Laws placed high duties on the import of wheat when its price at home was above 53s. 4d. a quarter (48s. from 1773), and gave a bounty of 5s. a quarter on wheat exported when its price at home was below 48s. (44s. from 1773). The prohibition on imports from Ireland must have improved the position of English graziers and dairymen, just as the duties on imported corn must have tended to raise the price of corn in short years, although in the earlier eighteenth century there was generally a surplus for export; and the 5s. bounty on export was undoubtedly a factor in encouraging the improvement of cultivation in favourably-situated areas such as East Anglia. The advantages of the government's paternal attitude towards farming were considerably modified, however, by the landlord's acceptance of heavy taxation of land in the numerous war years of the period. The land tax was not levied equitably, however, and it is important to notice that its real burden was greatly reduced as prices and rentals rose in the later eighteenth century. But in the south and east of the country, where the tax fell most heavily, the reduction of landowners' incomes by up to a fifth in the first half of the century must have had unfavourable effects on their investment in their estates, especially for the lesser gentry who had few or no sources of income beyond their rentals.

The legal developments which chiefly affected agriculture were concerned with the management of estates and the ability to borrow on the security of land. In the eighteenth and nineteenth centuries the

[1] G. E. Mingay, 'The Size of Farms in the Eighteenth Century' *Econ. Hist. Rev.* 2nd ser. XIV (1961–2).

majority of estates of any size, and something over a half of the cultivated land, were governed by the provisions of strict family settlements. Legal decisions in the closing years of the seventeenth century, and further ones in the eighteenth century, sanctioned the practice which lawyers had already adopted of so restricting the descent of property that it was always likely to be inherited intact by the eldest son of the landlord. But in order to safeguard the value of the inheritance considerable restrictions were placed on its use by tying the hands of the man presently in possession: he could not sell settled lands, his powers of borrowing were restricted, the income of certain of the settled lands might be reserved for particular purposes, such as paying for the dowries settled on the daughters of the family or the jointure of a widow, and there might also be rules relating to mining and the cutting of timber.

It is impossible to say how far these various limitations on the freedom of landowners affected the development of farming, but clearly they must have gone some way towards offsetting the advantages for enclosure and consolidation which the accumulation of land in large estates offered. However, the family settlement did not place the management of landed estates entirely in a strait-jacket. Not all of the land need be settled, and under certain circumstances the settlement could be set aside and disregarded. There was, in addition, a tendency for lawyers when drawing up these settlements to frame more liberally the clauses dealing with borrowing and the uses that could be made of the land.

Moreover, the borrowing powers of landowners were greatly enhanced by the development of the mortgage. At one time to mortgage land in order to borrow involved a very considerable risk that the land put in pledge might be forfeited, since the courts took a strict and unsympathetic attitude towards even very slight infringements of the terms of the mortgage, and to be one day late in payment, for example, might well involve the loss of the land in question. Here changes in the law worked in favour of greater freedom and security of borrowing, and so reduced the risk that large mortgages might lead to the break-up of an estate. Landowners of all kinds, from the great owners down to the small freeholders, widely availed themselves of the mortgage in order to buy more land, to enclose or otherwise improve the land they already had, or to build a more splendid residence, meet the cost of children's education and portions, or pay for election contests. Mortgages were arranged through London bankers or local attorneys in provincial towns, and as the capital market developed and lenders and borrowers were more readily brought together so the cost of borrowing was considerably eased, the rate of interest on mortgages and personal bonds falling to 5 per cent, or even as low as 4 per cent for some favoured borrowers.

Legal developments thus had conflicting effects: on the one hand family settlements became stricter and somewhat hampered landowners in the management of their property; on the other hand the changes in the legal position of the mortgage and the development of the capital market made it easier and cheaper to raise capital for estate expansion and improvement.[1]

Even when agricultural improvement was encouraged by the rising prices of the later eighteenth century, progress was not everywhere very rapid. In areas of suitable soils and good communications, as we have noticed, progress was much more remarkable than in less favoured areas. Some districts remote from good markets and lacking fertile loams or improveable sands, or having clays unsuited to leys or permanent pasture, tended to remain outside the mainstream of progressive farming. There were other factors, however. The structure of landownership might tend to the perpetuation of small and inconvenient farms and so delay enclosure almost indefinitely, as in parts of the Midlands and some other areas. Small farmers generally lacked the acreage and capital to undertake convertible husbandry or improve the quality of their stock and grasslands, and they tended to be ignorant and opposed to change. The conservatism of many landlords and larger farmers also slowed down the pace of change, although it must be remembered that caution was only prudence when many of the new practices and implements advocated by the experts were never practicable everywhere, nor always sensible.

Tithes may have been a factor of some importance, too. To Arthur Young, the tithe was 'the greatest burthen that yet remains on the agriculture of this kingdom; and if it was universally taken in kind, would be sufficient to damp all ideas of improvement.'[2] Farmers objected to giving away a tenth of their produce, and equally a tenth of any increase in yield which their skill and enterprise made possible. Their hostility to the payment of tithes was such, we are told, that they often went out of their way to annoy or cheat the tithe-holder. One farmer, it was said, once informed the parson that he was going to pull a field of turnips, and when the parson's man arrived with his cart the farmer pulled just ten roots, gave one to the carter and told him he would let his master know

---

[1] For a fuller discussion of these changes see Mingay, *English Landed Society* pp. 32–9.

[2] A. Young, *Political Arithmetic* (1774) p. 18.

when he would pull some more. Such stories undoubtedly exaggerated the difficulties over tithe, however. Although the practice varied both from place to place and from time to time, the fact was that tithes were very widely paid in cash rather than in kind, and where the landlord was the lay impropriator he merely added the value of the tithes to the rent.[1] Tithes were a burden and a nuisance, but hardly a real bar to improvement.

Even more exaggeration was given to the importance of leases. Young argued that 'the improvements which have been wrought in England have been almost totally owing to the custom of granting leases.'[2] Young, of course, was the great apostle of improvement, and like all enthusiasts and propagandists he tended to overstate his case. He was not alone in his opinions, however: many agricultural writers of his time agreed with him that security of tenure was vital to a farmer if he were to invest money in expensive and durable improvements and take a long-term view of the potentialities of his land. A long lease of 14 or 21 years, renewable perhaps every seven years, was considered ideal by some writers, but others favoured shorter terms as better for safeguarding the interests of both farmer and landlord.[3]

What is known about the actual use of leases, however, does not accord with the assertions of contemporary authorities. In the early part of the century especially, there was a tendency for the old-fashioned and unsatisfactory copyhold tenure and lease for lives to be replaced by leases for definite terms of years, but where such ancient tenures did not prevail many of the smaller farms, and even some large ones, were let not on lease but on annual agreements. Yet their occupiers rarely felt insecure; they trusted that the landlord would follow the accustomed rule of leaving occupiers in possession so long as they proved satisfactory tenants. Thus we hear of these tenants occupying farms for long periods on annual tenures, even passing them on to their sons, widows or daughters, and often investing in improvements. Leases were certainly more common with larger farms, but mainly because the value of the property was such that both landlord and tenant felt the need for a formal definition of their respective rights and duties.

It does not appear that leases containing detailed covenants governing the actual mode of farming the land were very common (apart from the normal inclusion of the conventional restrictions against breaking up

[1] See W. Marshall, *Rural Economy of the Midland Counties* (1796) p. 18.
[2] Young, *op. cit.* p. 15.
[3] See W. Marshall, *On Landed Property* (1804) pp. 362–5.

permanent pasture and selling manure off the farm), and those having covenants of a definitely progressive nature even less so. Professor Habakkuk is of the opinion that leases probably made a negative rather than a positive contribution to agricultural progress, by the leaving out of old covenants designed to maintain traditional practices, rather than by the inclusion of covenants to introduce or enforce new methods.[1] On the Coke estates, however, some of the leases made in the 1730s were remarkable for their insistence on a two-year ley, and in the second half of the century the husbandry covenants were enlarged and improved. Under the famous Coke of Norfolk the husbandry covenants ceased to be negative prohibitions and became positive instructions which bound the tenant to endeavour to follow a prescribed six-course rotation.[2] It should be remembered, however, that this was 'a highly disciplined and ordered estate', presided over by a renowned pioneer among landlords and a famous estate steward, and its standards were almost certainly far above those of the great majority of even the large estates. But even where progressive covenants were in use it is difficult to say how strictly they were enforced, or whether the farmers might not have done as well without them. And it is clear from estate records and contemporary literature that some landlords persisted with outdated husbandry covenants which unnecessarily tied the tenant's hands and prevented experiment.

Again, long leases tended to decline in popularity in periods of rapidly changing prices, for then the long lease was bound to favour one party at the expense of the other: in a period of rising prices the landlord found his farms under-rented and was incapable of raising rents until the leases fell in; in a period of falling prices the tenant found himself bound to pay a rent that had seemed reasonable when prices were higher but which he could no longer afford. Possibly the tenants had the better part of the bargain for if prices fell severely they might force the landlord to make temporary abatements under threat of giving up their farms, while it was unknown for landlords to raise rents during the period of a lease when prices were rising. In the eighteenth century the price changes were perhaps not sufficiently rapid or variable to affect very much the use of leases, but the long and sharp price climb after 1793, followed as it was by the post-1815 depression, seems to have left both landlords and tenants much more wary of leases.

[1] H. J. Habakkuk, 'Economic Functions of Landowners in the Seventeenth and Eighteenth Centuries' *Explorations in Entrepreneurial History* VI (1952) p. 93.
[2] R. A. C. Parker, 'Coke of Norfolk and the Agricultural Revolution' *Econ. Hist. Rev.* 2nd ser. VIII (1955-6) pp. 159-63.

Lastly, it is likely that some landlords preferred not to grant leases in order that they might have a closer control over those of their tenants who were entitled to vote at the county elections. But probably only a small minority of landlords were prepared to keep their tenants in a state of constant insecurity for political motives, refusing them leases and depriving them of the conventional security enjoyed by the majority of annual tenants. Such a policy was bound to affect the farming of the estate, and before long the landlord would be obliged to recognize its effects on his rentals. Except therefore for its use in areas of large-scale and progressive farming like Norfolk, it does not seem that in general we can regard the lease as a very important instrument in agricultural improvement, nor its absence as a great obstacle to efficiency. Its most valuable role was probably in encouraging farmers to undertake the risks of cultivating newly-enclosed waste lands by guaranteeing them low rents for the early years of the tenancy. Even here, however, we need more knowledge of its actual use before we can be sure just how important it really was.

As yet we have said little about the feature of eighteenth-century farming which is usually thought to have constituted the main obstacle to improvement—the open fields. Young, of course, condemned outright everything about them, deploring the ignorant practices of their 'goths and vandals' of farmers, and claiming that 'without inclosures there can be no good husbandry'.[1] Reporting on Oxfordshire for the Board of Agriculture, he commented pungently on the gulf between the new men with enclosed farms and progressive ideas, and the diehard conservatism of the old type of open-field husbandmen: 'When I passed from the conversation of the farmers I was recommended to call on, to that of men whom chance threw in my way, I seemed to have lost a century in time, or to have moved a thousand miles in a day. Liberal communication, the result of enlarged ideas, was contrasted with a dark ignorance under the covert of wise suspicion; a sullen reserve lest landlords should be rendered too knowing; and false information given under the hope that it might deceive. . . . The old open-field school must die off before new ideas can become generally rooted.'[2]

As usual Young overstated his case, but undoubtedly the efficiency of open-field farmers often suffered from the survival of old methods and conditions which they were unable or unwilling to change. These included

[1] Young, op. cit. p. 198.
[2] A. Young, General View of the Agriculture of Oxfordshire (1809) pp. 35–6.

the dispersal and fragmentation of the holdings and the time wasted in journeying with implements from one part of the field to another; the unimproved nature of the soil, and the waste of land in balks (although these served as additional pieces of pasture as well as paths between lands and headlands for turning the plough[1]); the rigid rotation of two crops and a fallow; the impossibility of improving the livestock, and the risks of wildfire spread of disease among beasts herded together on the commons and fields—all remained serious drawbacks. We hear of villages on heavy clay land (for example in the Holderness region of the East Riding) where a two-field system, one year in crop and one year in fallow, still persisted, unaltered perhaps from medieval times, although a three-course rotation was much more common; of villages where sheep rot and the cattle plague periodically decimated the low-grade, nondescript beasts that made up the village flocks and herds (an Oxfordshire farmer said 'I have known years when not a single sheep kept in open fields escaped the rot. Some years within living memory rot has killed more sheep than the butchers have'[2]); and we hear of farmers like Richard Derby of Hanslope in Buckinghamshire, whose 26½ acres of open-field land was spread over three fields in 24 parcels, some as small as a quarter of an acre.[3]

Perhaps the most striking weakness of the system, however, was the annual fallowing of a proportion, generally from a quarter to a third, of the arable land. This was necessary in order to restore fertility after two or three years of cropping, and further by successive ploughings to break up the surface thoroughly and get rid of the accumulation of weeds. Fallowing was essential, and indeed remained so on heavy, wet clays that were enclosed, but for more amenable soils that could be brought within the scheme of convertible husbandry it represented an unnecessary loss of output and unduly high average costs of production.

Nevertheless, it is now appreciated by historians of the period that open-field farming was far from moribund, and indeed showed many clear signs of vitality. There was in many villages a growth of small enclosures, an increase in the number of fields, greater flexibility of rotations and the cultivation of the formerly fallow field. The open fields

[1] See the discussion by H. A. Beecham, 'A Review of Balks as Strip Boundaries in the Open Fields' *Ag. Hist. Rev.* IV (1956) pp. 22–44.

[2] A. Young, *General View of Oxfordshire* (1809) p. 102. See also T. Stone, *Lincolnshire* (1794) p. 62; J. Crutchley, *Rutland* (1794) pp. 14–15; C. Vancouver, *Cambridgeshire* (1794) pp. 87, 208; T. Rudge, *Gloucestershire* (1807) p. 250; G. Rennie, etc., *Yorkshire* (1794) p. 32.

[3] Nottingham University, Manvers Colln: Survey of Hanslope 1763.

were never (as Professor Tawney ironically described them) a 'perverse miracle of squalid petrifaction', nor their cultivators 'the slaves of organized torpor.'

Open-field farming could be made much more adaptable and efficient if the open arable was amply supported by pasture closes which might be used for rearing, dairying and fattening. There was in fact a strong tendency for these valuable closes to grow in number (a development stimulated perhaps by low corn prices in the first half of the century), the farmers agreeing to take land out of the open fields for the purpose. The process may be readily observed in the accounts of the Duke of Kingston's estates in the 1730s and 1740s: At Eakring in Nottinghamshire 35 acres were taken out of the Grass field between 1744 and 1746, divided into pasture closes and added to the farms of some of the bigger farmers, and at Crowle in the Isle of Axholme there were three new enclosures in 1732, while at Hanslope the Willow Hedge field was enclosed in 1734.[1]

In consequence the open fields themselves tended to shrink in area, and in some villages still open at the end of the eighteenth century there was even as much or more land in closes as in the fields, a very significant modification of the system. The growth of small closes of course goes back long before our period. In 1691 half of the land of the well-known village of Laxton (whose open fields are still worked today), as many as 950 acres out of 1,908 acres, already lay in closes. A hundred years later nearly 70 per cent of the total area was in closes, and the proportion of land devoted to arable had consequently fallen to as little as 31 per cent. In the dairying village of Hanslope in Buckinghamshire, too, most of the land was enclosed before 1763. In that year the village contained 1,099 acres of enclosures (of which only 189 acres were enclosed arable and the remainder were enclosed pasture and meadow), 578 acres of open arable, and 85 acres of open meadow.[2]

In the open fields themselves there were often exchanges of lands to obtain more compact and convenient holdings, and arrangements to allow more advanced rotations to be adopted. The open-field byelaws were not inflexible, but were frequently amended and supplemented, and it should be noticed that the actual rotation in use was not necessarily governed by the number of fields. Some two-field villages in Holderness, for instance, had in fact a three-course rotation, the cropped field being sown half with

[1] Nottingham University, Manvers Colln.: Kingston Estate Accounts.
[2] B. M. Egerton MS. 3564; Nottingham University, Manvers Colln.: Kingston Surveys.

winter grains—wheat and rye—and half with spring crops—barley, roots, peas and beans.[1]

The desire of open-field farmers to improve their livestock management and carry more stock was similarly reflected in the increase in enclosed meadows and the stinting or rationing of the common grazing lands. In Nottinghamshire again, the farmers of Cotgrave agreed in 1717 to enclose one of their four open fields in order to remedy a shortage of grass land, and rules were drawn up for sharing and maintaining the new pasture. At Beighton in Derbyshire a Mr Samuel Buck was paid £44 'for drawing several writings with the Freeholders of Beighton concerning the new Enclosure and several Exchanges of Land, and the payment of annual sums of money in lieu of Tythe Corn.' And in Hampshire the farmers of West Grimstead turned their common down into a fourth open field, 'so that they have alwaies one Grass Field for their Sheep, and they find so much Convenience and Advantage by it, that the Tenants of East Grimstead are very eager and importunate to do the same'—so much so, reported the landlord's steward, that they were willing to pay a heavy fine of 40s. per acre for permission.[2]

In the actual cultivation of the land the open-field farmers also showed enterprise. Fallowing was reduced, and there was the introduction, in Oxfordshire for example, first of sainfoin and later of turnips. The sainfoin, more nourishing than ordinary grass or hay, was particularly valuable in feeding larger flocks and herds, and so in providing a larger supply of manure for the arable crops. Where there were only two open fields they were sometimes divided to make three or four, one field being sown with sainfoin.[3] Leys, i.e., lands laid down with legumes like sainfoin or clover or with improved grasses like ryegrass, and then ploughed up after a few years for corn crops, were adopted in both open fields and closes to provide additional pasture, to reduce bare fallows, and restore fertility. Through more closes and open-field leys farmers could overcome, to a considerable extent, the shortage of pasture which formerly limited the number of animals they could keep, and hence enabled them to diversify their farming and keep a reduced area of arable in better heart. Dr Hoskins found this 'evidence of a convertible husbandry in the open fields [goes] well back into the sixteenth century, and probably earlier

[1] A. Harris, *The Open Fields of East Yorkshire* (East Yorks. Local History Series No. 9, 1959) pp. 4–6, 22.

[2] B. M. Egerton MS. 3622 ff. 202–4; Nottingham University, Manvers Colln.: Kingston Accounts; Lincoln R.O.; Monson S.C. XCIII.

[3] M. A. Havinden, 'Agricultural Progress in Open-Field Oxfordshire' *Ag. Hist. Rev.* IX (1961) pp. 74–82.

than that.' In the seventeenth century the larger open-field farmers of Leicestershire extended leys up to a quarter or nearly a third of their acreage, and sometimes the leys would be permanently enclosed, a factor in the process of slow, piecemeal enclosure of open-field villages. Furthermore, Dr Hoskins has exploded the fallacy that open-field farmers were strictly bound by traditional rotations. In the seventeenth century, the Leicestershire farmer kept no strict proportions between crops, and 'within certain broad limits could do what he liked with his own strips.'[1]

This and much other evidence of the cultivation of turnips and artificial grasses, the use of water meadows for a bite of early grass, and systematic folding of the arable by large communal flocks under a common shepherd, all show that the ancient structure was not so backward nor so incapable of improvement as was once supposed. But as it was modified and improved the old system was gradually stripped of its distinctive characteristics, and moved towards the individual control of the land, the freedom in land use, and the compact and larger farm units associated with enclosed farms. Left to itself the development of open fields would no doubt have arrived in the fullness of time at completely enclosed and individually managed farms. Enclosures by agreement between the owners and by Act of Parliament speeded up the process. It follows that many enclosures merely completed the work of centuries, and made little change of significance in the farms, their occupiers or their methods of cultivation. The view that *all* enclosures represented a sudden breach with tradition, and brought about a transformation of the farming system and social structure of the village, has no foundation in fact.

Nevertheless, enclosure was necessary because not all open-field villages showed much progress or efficiency, and because even where there was progress there were limits to the profitability of the land so long as even the truncated open fields and commons remained a part of the farming system. Enclosure was the means of bringing about rapidly a large increase in the cultivated area and in the productivity of the soil as the responses to prices and market conditions. The best use of the land could be obtained only in enclosed and reasonably large farms. The farmers themselves recognized this by their willingness to pay something like twice as much as the open-field rent for the same land in enclosures. However, before we go on to examine the enclosure movement in detail

[1] W. G. Hoskins, 'The Leicestershire Farmer in the Seventeenth Century' *Agricultural History* XXV (1951) pp. 14–17, 20.

we must say more about the advanced farming practices which progressive farmers were adopting at this time.

*References and suggestions for further reading:*

| | |
|---|---|
| T. S. Ashton | *An Economic History of England: the Eighteenth Century* (1955), Ch. II. |
| J. D. Chambers | *The Vale of Trent* (Supplement No. 3 to the *Economic History Review*). |
| G. E. Fussell | 'The Size of Cattle in the Eighteenth Century' *Agricultural History*, III (1929). |
| W. G. Hoskins | 'The Leicestershire Farmer in the Seventeenth Century' *Agricultural History* XXV (1951), or *Provincial England: essays in social and economic history* (1963). |
| A. H. John | 'The Course of Agricultural Change, 1660–1760' in *Studies in the Industrial Revolution* (ed. L. S. Pressnell, 1960). |
| E. L. Jones | *Seasons and Prices* (1964). |
| E. Kerridge | *The Agricultural Revolution* (1967). |
| G. E. Mingay | *English Landed Society in the Eighteenth Century* (1963), Ch. VII. |
| G. E. Mingay | 'The Agricultural Depression, 1730–1750' *Economic History Review* VIII (1955–6). |
| G. E. Mingay | 'The Size of Farms in the Eighteenth Century' *Economic History Review* XIV (1961–2). |
| R. A. C. Parker | 'Coke of Norfolk and the Agricultural Revolution' *Economic History Review* VIII (1955–6). |

# 3 The New Farming

The fundamental improvement in eighteenth-century farming consisted in the spread of more flexible rotations of crops, embracing roots, legumes and improved grasses which by providing more fodder enabled the land to carry more stock, which in turn enriched the soil with their manure. Heavier manuring raised the yields of both cereal and fodder crops and hence made possible even heavier stocking—a vital advance for a farming system based on land which had been cultivated for many centuries and which lacked efficient artificial fertilizers. More land and more suitable conditions for the new crops and larger numbers of animals were provided by enclosure of open fields and commons and the taking in of wastes, but the growing adoption of alternate husbandry—the famous 'Norfolk system'—on the light soils, and of convertible husbandry—'ley-farming' —on the heavier but still sufficiently well-drained lands, represented the real breakthrough in farming techniques. The eighteenth century saw many other changes and developments: the improvement of livestock by selective breeding, experiments in drainage and the treatment of the soil with marl, chalk, bones and other materials, and the introduction of new machinery and better-designed implements. These were also important, but rather for what they promised in the future than for their present advantages.

The central advance in rotations and heavier stocking brought many significant gains over the ancient succession of one or two corn crops and a fallow. First, the cultivation as field crops of roots like turnips, and later swedes and mangolds, and of rich legumes and grasses like clover, sainfoin and ryegrass, provided great quantities of additional fodder and raised the levels of animal nutrition. This was vastly important for improving the output and quality of meat and dairy produce. Formerly, inadequate nutrition had meant that beasts were long and expensive in fattening for the market, produced low yields of milk, and were unable to achieve their potential fertility. The old belief that shortage of winter fodder necessitated a heavy autumn slaughtering is undoubtedly exaggerated, but certainly the lack of feed kept down the numbers of stock that could be carried and so restricted the output of animal products.[1] Moreover, the mixed grazing of herds and flocks on the commons, hill-pastures and open-field stubbles, often very ill-drained, hampered the development of improved breeds and facilitated the spread of the murrains, rots and other afflictions.

It is important to notice that of the new fodder crops turnips, for all their textbook fame, were the least in nutritive value as fodder, and they spread more slowly than did the richer legumes and grasses. The importance of turnips lay more in the flexibility they gave to arable rotations on light soils, and the scope they offered for manuring the soil when fed off on the ground to sheep in winter. On heavier soils turnips might still be grown if the ground was not too wet, but mainly, as in the western counties of Hereford, Devon and Cornwall, in small quantities as a supplement to grass for feeding in yards or byres. In the south-west grass grew all the year round and turnips were not therefore vital, while on heavy lands generally turnips had to be harvested since sheep had the effect of compacting the soil and could not be wintered on the arable fields. Another serious and little-appreciated limitation of the turnip was its vulnerability to certain weather conditions. In a winter which brought an alternation of sharp frosts and rapid thaws turnips were liable to rot in the ground; in a mild spring they were likely to run to seed; and in a dry summer they cropped but thinly. The increased dependence on the turnip for fodder which became a marked feature of light soil areas in the eighteenth century was therefore dangerous, for it meant periodical severe fodder shortages and forced reductions in flocks and herds in seasons when the turnips failed. There consequently developed a search for alternative and more reliable substitutes, which eventually appeared in the form of the hardier swede (Swedish turnip), introduced in the later years of the century, kale, early clover and vetches, and in the middle years of the nineteenth century the mangel-wurzel, a root that could be grown on heavy soils where turnips would not thrive.[2]

[1] See R. Trow-Smith, *A History of British Livestock Husbandry to 1700* (1957) pp. 255–9.

[2] See E. L. Jones, 'Agricultural Conditions and Changes in Herefordshire, 1660–1815' *Trans. Woolhope Naturalists' Field Club*, Vol. 37 (1961) pp. 38–40; 'English Farming before and during the Nineteenth Century' *Econ. Hist. Rev.* 2nd ser. XV (1962–3) p. 146; *Seasons and Prices* (1964) pp. 93–6, 121–4.

The turnip was first introduced from Holland as a vegetable in the later sixteenth century, and by the middle seventeenth century was in widespread use as a field crop for fodder in Norfolk and Suffolk and other parts of eastern and southern England.[1] The legumes such as clover, sainfoin, trefoil and lucerne, and the cultivated or 'artificial' grasses such as ryegrass, cocksfoot, and meadow fescue, were used both for grazing stock and for cutting for hay. Clovers were sown under barley in arable rotations and fed off to stock in the winter after the barley was harvested, while the other new crops were usually sown as leys or temporary pastures and ploughed up after one, two or three years. Sainfoin and clover were both widely established, even in open-field areas, by the middle or later seventeenth century, and sainfoin and lucerne in particular provided a highly nutritious feed, twice as rich as the ordinary native grasses.[2] Another important factor in the increased supply of feed was the spread of the water-meadow through the west country and the west Midlands during the later sixteenth and the seventeenth centuries. Streams were dammed to spread a thin but constant flow of water over the meadows, and by this means a rich and early bite of grass was provided,[3] most valuable for meeting the 'hungry gap' of March and April when the spring growth of grass in the ordinary pastures might be held back by cold weather.

A vital consequence of the heavier stocking made possible by the improved supply and quality of fodder was the enrichment of the soil. This was particularly important on the light soils which could only bear good crops of corn when heavily manured by the beasts wintered on the turnips, clovers and sainfoin. Furthermore, the actual cultivation of the roots, legumes and grasses helped to improve the soil. Although eighteenth-century farmers did not understand the peculiar power of clover to fix atmospheric nitrogen, they appreciated its mysterious value in restoring soil fertility after the exhausting crops of wheat and barley, and they found too that the lands benefited from the ploughing in of the leys that had served their turn; and if they had read their Jethro Tull they knew that the systematic hoeing of roots sown in drills produced a fine tilth which helped the young plants to thrive, and at the same time kept down the weeds. Under these systems of alternate or convertible husbandry

---

[1] E. Kerridge, 'Turnip Husbandry in High Suffolk' *Econ. Hist. Rev.* 2nd ser. VIII (1955–6) p. 390.

[2] Trow-Smith, *op. cit.* pp. 255–7.

[3] E. Kerridge, 'The Sheepfold in Wiltshire and the Floating of the Water-meadows' *Econ. Hist. Rev.* 2nd ser. VI (1953–4) pp. 286–9.

the land was always growing something of value—if not a cash crop of corn then valuable fodder for the animals—without exhausting its fertility, and so the wasteful bare fallows were virtually abolished. Lastly, the larger numbers of beasts kept by farmers practising the new husbandry made them rather less dependent financially on the marketing of corn crops and gave them a more diversified output.

Of course, the improved cropping systems were not adopted everywhere, if only because of the conservatism of the farmers and their resistance to innovation, a state of mind often reinforced by the traditional restrictions and unsuitably small holdings of open-field farming. Contemporary writers frequently attacked the prevalence of small farms as a factor responsible for the slow spread of improved husbandry, and Edward Laurence, John Arbuthnot, Arthur Young and many of the Board of Agriculture's county reporters were among those who believed that the small farmer, with his ignorance, traditional outlook and lack of capital, was always a bad farmer. 'Great farms', said Young, 'are the soul of the Norfolk culture.' There was a tendency to think that the 'spirit of improvement', in Marshall's phrase, was less evident in the western and northern extremities of the country than in the eastern counties, the Midlands and south.

But a more valid division, cutting across regions, was between the free-draining soils of the better sands in Norfolk and the chalk and limestone uplands of the Downs and Cotswolds, and the heavy, wet soils of the midland plain and clay vales. The first offered scope for improvement and saw many changes in cultivation. In the Hampshire sheep-and-corn country, for example, the introduction of the new fodder crops enabled the farmers to reduce their stock, formerly invaluable for keeping the arable in heart, and extend their arable cultivation into the down sheep-runs.[1] But in the Vale of Belvoir after enclosure in the later eighteenth century, the former rich open-field arable of the valley bottom was put down to grass, while commons were ploughed up and cultivation extended up the poorer land and sheepwalks of the hillsides, the total area of arable being considerably reduced in favour of grass.[2] In general, the extension of grain production to the free-draining uplands not only expanded the total area of arable cultivation, but also supplemented the

[1] E. L. Jones, 'English Farming' *loc. cit.* p. 146; 'Eighteenth-century Changes in Hampshire Chalkland Farming' *Ag. Hist. Rev.* VIII, 1 (1960) pp. 8–15.

[2] W. G. Hoskins, 'The Leicestershire Crop Returns of 1801' in *Studies in Leicestershire Agrarian History*, ed. Hoskins (Trans. of Leics. Arch. Soc. XXIV, 1948) pp. 131–2.

traditional granary of the clayland vales in a valuable way, for good crops might still be grown on the light soils in seasons that were too wet for the clays.[1]

The difficulty of growing roots and the new legumes and grasses on the other main type of land, the wet and cold clays, provided a serious and persistent obstacle to agricultural progress in the midland clay triangle and other districts of similar soils. The existing methods of drainage do not seem to have been effective in improving the adaptability of these soils: they remained too wet for the Norfolk system, and too wet also to be really safe as leys or permanent pasture. Of necessity, two crops and a fallow remained the basic rotation on them, even when the land was enclosed, until cheap under-drainage came in towards the middle of the nineteenth century. And because the wet clays were so difficult to improve and so expensive to cultivate, landlords and large, progressive farmers tended to avoid them and listened to Arthur Young's advice: 'such soils must, like others, be cultivated by somebody, but I would advise every friend of mine to have nothing to do with them.'[2]

The unfortunate cultivators of these soils were indeed obliged to meet heavy expenses of cultivation: to employ large teams which could drag the implements through the sticky soil for as many as six ploughings and five harrowings in order to produce a seed bed; to see the teams stand idle in wet weather because the ground was unworkable; and to find at the end of it all that their yield was lower and their profits less than could be obtained on the more easily-worked soils. 'The farmers have a proverb among them, which seems a very true one', said Arthur Young: '*a man cannot pay too much for good land, nor too little for bad . . .*'. On cold, hungry, wet soils, he went on, 'the lower sort of farmers fallow them for wheat; and then take a crop of oats, and fallow again; or use them in some such unprofitable course, as all that can, without improvement, be made of them. Such soils I should strenuously advise any man from hiring, however low the rent may be, unless for improvement. It is impossible to calculate the produce of such soils; scarce one season in twenty suits them; in wet years they are nothing but mud, and yield nothing but weeds; in dry ones, they bake with the sun after rain, so that the corn is bound into the ground; it is only middling years peculiarly favourable, that can permit these soils to bear a tolerable crop. Now such lands . . . are so extremely perplexing to manage, so tedious in every operation, and so

---

[1] E. L. Jones, *Seasons and Prices* (1964) pp. 110–11.
[2] A. Young, *The Farmer's Guide in Hiring and Stocking Farms* (1770) II p. 4.

particularly *late*, that a man had better hang himself than have anything to do with them.'[1]

However, wherever the soils were sufficiently well drained, and shortages of winter fodder and naturally thin and infertile soils made the high labour costs of the Norfolk system worth while—in the home counties, East Anglia, and much of southern England—the Norfolk four-course rotation of wheat-turnips-barley-clover, or some variation of it, was well established even in the late seventeenth century. William Marshall, writing in 1787, recognized this when he said that the farms of north-east Norfolk 'have been kept invariably, *for at least a century past*, under the following course of cultivation: wheat, barley, turnips, barley, clover, ryegrass' (our italics).[2] On the estates of the Coke family in north-west Norfolk the use of alternate husbandry can be traced back in the records to at least the 1720s, and on the neighbouring estates of the Walpoles to 1673.[3] As we have seen, it is known that both turnips and clover were widely grown on light soils in the seventeenth century, and it is recognized that there was a strong connection between the growth of commercial farming in Holland and the introduction of these new crops to England in the sixteenth and seventeenth centuries. The Flemish farmers had steadily developed the intensive cultivation of their poor, sandy soils since the Middle Ages, introducing fodder crops and heavy manuring in order to obviate fallows, and using bed and row cultivation to obtain high yields and make their small farms commercially viable. It was not surprising that clover, sainfoin and ryegrass should migrate across the North Sea to where the English farmers found themselves faced with similar problems and wished to farm for expanding markets. In both Holland and in England the periods of low corn prices in the seventeenth and early eighteenth centuries saw the extension of more intensive methods of farming, and the conversion of some lands from arable to pasture.[4]

It follows from all this, of course, that those celebrated eighteenth-century pioneers, Tull, Townshend and Coke of Norfolk, were more the popularizers of the new methods of intensive cultivation than the originators

---

[1] A. Young, *Rural Economy* (Dublin, 1770) pp. 149, 154.

[2] W. Marshall, *Rural Economy of Norfolk* (1787) I p. 132.

[3] R. A. C. Parker, 'Coke of Norfolk and the Agrarian Revolution' *Econ. Hist. Rev.* 2nd ser., VIII (1955–6) p. 160; J. H. Plumb, 'Sir Robert Walpole and Norfolk Husbandry' *ibid.* V (1952–3) p. 86; G. E. Fussell, in Lord Ernle, *English Farming Past and Present* (6th ed. 1961) Introduction, p. lxvii.

[4] B. H. Slicher Van Bath, 'The Rise of Intensive Husbandry in the Low Countries,' in *Britain and the Netherlands* (ed. J. S. Bromley and E. H. Kossmann, 1960) pp. 132, 139, 142, 150.

of them. They drew on the long-established commercial farming methods of the Dutch, and even on English predecessors like Sir Richard Weston, who in the seventeenth century did much to forward the growing of fodder crops in leys: In 1645 Weston published the first account of the Low Countries' farming, although even before this Flemish immigrants and English visitors to the Low Countries had introduced market gardens on the Dutch pattern, and root crops, legumes and artificial grasses also.[1] Tull, indeed, had many original ideas, particularly in the designing of horse-drawn implements for sowing in drills and for hoeing. But even here he was preceded by Worlidge, who planned a practical machine some sixty years earlier, while Tull's emphasis on sowing in drills rather than broadcast, and on hoeing systematically to produce a fine tilth and to destroy weeds, were also not entirely new.[2] And of course his view that such methods made it possible to dispense with crop rotations and manures (which he considered to be of little advantage in improving fertility), was misleading. But he was not quite the dogmatic crank that some writers have made him out. He recognized that his system was not applicable to all conditions. It was inappropriate, he said himself, for 'wet clayey land. . . . Many, it is like, will think, this repetition of wheat crops rather a curiosity than profitable, and in some circumstances it may be so.'[3] On the whole, we cannot regard Tull as a vital innovator, but his work well illustrates the spirit of enquiry and interest in mechanical contrivance which were characteristic of the 'spirited improvers' of the age.

Of Townshend we cannot with certainty say even as much. He may well have helped to popularize the turnip as a field crop but it was certainly firmly established in the Norfolk rotations when he was born, as was also the practice of marling, so important for the thin sandy soils of the area, which he was once thought to have revived. It used to be supposed that Townshend's property when he inherited it was little more than a sandy waste, 'where a few sheep starved and two rabbits struggled for every blade of grass', and that Townshend transformed it by liberal use of marl and by cultivation on the Norfolk system.[4] But when the full facts are known they will probably show that Townshend, like Coke, merely developed existing practices and improved, rather than transformed, his

[1] G. E. Fussell, 'Low Countries' Influence on English Farming' *Eng. Hist. Rev.* LXXIV (1959) pp. 611-16.

[2] G. E. Fussell, *The Farmer's Tools* (1952) pp. 93-102.

[3] E. R. Wicker, 'A Note on Jethro Tull: Innovator or Crank?' *Agricultural History* (1957) pp. 47-8.

[4] Ernle, *op. cit.* pp. 173-5.

property. Coke of Norfolk, of course, lived at a considerably later period. He inherited his estate only in 1776, and his death came a full hundred years after the deaths of Tull and Townshend. Coke was an important innovator, but not in the way that has usually been supposed. He improved the content of his leases, he was instrumental in improving the breeds of sheep and cattle in Norfolk, he encouraged irrigation, underdrainage and the use of new manures, and not least, his 'sheep-shearings' or private agricultural shows probably did much to spread the knowledge of improved farming. But he did not introduce or transform the main features of Norfolk farming: the large farms, long leases, the Norfolk rotation and marling were all well established on his estates before his time.[1] That Bakewell, the fourth of the great eighteenth-century names, needs also some re-assessment we shall see in a moment: but already it is clear that we can no longer accept the heroic view of an agricultural revolution springing from the originality and enterprise of a mere handful of great improvers. At the beginning of the eighteenth century improvement was already much more widespread than we once believed, and its origins go back far deeper into the past.

Over much of the midland counties the system of keeping land in leys had been practised since at least the sixteenth century. Ley-farming or convertible husbandry was found in the open fields as we have noted already, but the full advantages of the system were better reaped in enclosed and fairly large farms. Fundamentally, convertible husbandry and alternate husbandry (the 'Norfolk system') were closely related, for both relied on the alternate use of fodder crops and corn crops to obviate fallowing, keep the soil in heart and provide feed for the beasts. The differences were that in ley-farming there was usually no root break (although turnips were sometimes used to clean the soil of weeds if it was light enough), while the pasture (grass, clover or sainfoin) was kept down for a number of years, until it had deteriorated in quality and the land was again ready to bear three or so corn crops in succession. Even in the Norfolk system, however, the clover might be left down for two or more years as a ley, and so both systems were really variations of a basic crop rotation adapted to the needs of particular soils and climates. Marshall in his *Rural Economy of the Midland Counties* (1790)[2] succinctly explained the midland convertible husbandry: 'The outlines of management consist in keeping the land in *grass* and *corn* alternately . . . and in applying the grass to the breeding of heifers for the dairy, to *dairying*, and

[1] See R. A. C. Parker, *op. cit.* pp. 156–66.
[2] Vol. I pp. 184, 187.

to the *grazing* of barren and aged cows; with a mixture of ewes and lambs for the butcher. .... The land having lain six or seven years in a state of SWARD—provincially 'Turf'—it is broken up, by a single plowing, for OATS, the oats stubble plowed two or three times for WHEAT; and the wheat stubble winterfallowed for BARLEY and GRASS SEEDS;—letting the land lie, during another period of six or seven years, in HERBAGE; and then, again, breaking it up, for the same singular SUCCESSION OF ARABLE CROPS.'

Closely associated with both convertible and alternate husbandry was the improvement of the soil's texture and richness by marling, manuring and drainage. Thin sandy soils and impoverished upland ones greatly benefited from a dressing of a clay marl which gave them body and enabled them to retain water and manure near the surface, while unproductive clay soils were often improved by applications of chalk and lime. The action of chalk and lime was to break down heavy clay soils to a finer texture and make their natural fertility more readily absorbed by plants; and by reducing the tendency to stickiness in wet weather and to bake into hard lumps in dry seasons to make them easier to cultivate. Marl had the contrary effect of binding thin and sandy soils, as an early sixteenth-century historian of Pembrokeshire made clear:

> Claye marle is of nature fat, tough and clammy. The common people are of opinion that this marle is the fatnesse of the earthe, gathered together at Noah's flood: which is verie like to be true. .... It is digged or caste out of the pitte, carried to the lande, and there caste either upon the fallow or ley ground unplowed, and this in the summer tyme . . . where it lyeth so on the lande all the somer and winter, the rain making it to melte and run like molten ledd all over the face of the earthe.[1]

Marling had died out and been revived more than once since its earliest known use in prehistoric times, and it remained traditional practice in some counties as in Lancashire, Cheshire and Stafford; but it appears that the practice was greatly revived in the seventeenth and eighteenth centuries where landlords and tenants appreciated its advantages and were prepared to meet the cost of the work. By the eighteenth century marl was considered essential for the sands of Norfolk and Suffolk, and chalk brought across the Thames or into the river estuaries of East Anglia by boat from Kentish quarries was spread on the clays—the heavily-loaded wagons making a quagmire of the roads thereabouts. The im-

---

[1] Quoted by R. Trow-Smith, *English Husbandry* (1951) pp. 100–1.

portance of this for the farming of the region was not missed by Defoe, who noticed that the chalk was

> bought and fetch'd away by lighters and hoys, and carry'd to all the ports and creeks in the opposite county of Essex, and even to Suffolk and Norfolk, and sold there to the country farmers to lay upon their land, and that in prodigious quantities; and so is valued by the farmers of those countries, that they not only give from two shillings and six pence, to four shillings a load for it, according to the distance the place is from the said chalk-cliff, but, they fetch it by land-carriage ten miles, nay fifteen miles, up into the country. . . . Thus the barren soil of Kent, for such the chalky grounds are esteem'd, make the Essex lands rich and fruitful, and the mixture of earth forms a composition, which out of two barren extreams, makes one prolifick medium; the strong clay of Essex and Suffolk is made fruitful by the soft meliorating melting chalk of Kent which fattens and enriches it.[1]

Marl was dug in pits where it occurred near the surface and was carried in wagons to the fields, to be spread at the rate of about 20 wagon loads to the acre. If the marl was near at hand and had to be transported only a mile or so the cost worked out at about 40s. per acre.[2] This was a considerable outlay, which landlords were sometimes prepared to meet themselves, or for which they reduced the rent if the farmer were prepared to undertake it. Sometimes a farmer was bound by his lease to marl his land at so many acres per annum, thus spreading the cost over the whole term of the lease. The effects lasted for a good many years, and when the farms were let on 21 years leases the land was probably freshly marled with every new lease.[3] The proper amount of marl to lay on required good judgment, however, and the decline in the popularity of the practice in the nineteenth century may have been due in part to the harm done by excessive use of marl, as in Staffordshire.

An enormous variety of other manures were in use, from pigeon's dung and oxblood to soot and bones—the last leading to an appreciation of the value of bone meal. The most prized material, of course, was the farmyard manure, and one of the most heinous crimes of the unscrupulous tenant-farmer was to sell dung off the farm or to put it all on whatever land he owned himself. In the chalk and limestone uplands large flocks of sheep grazed the hills in the daytime and were folded every night on the arable land down in the vales. The folding was systematic, acre by acre, and often

[1] Defoe, *op. cit.* I pp. 99–100.
[2] See Marshall, *Norfolk* I pp. 150–7.
[3] Parker, *op. cit.* pp. 164–5; A. Young, *Southern Tour* (1768) p. 22.

the farmers pooled their flocks and hired a shepherd to regulate the business. Sheep there were valued more for their dung than their wool, and there was even evolved a type of sheep that very conveniently dropped its manure only at night while in the fold![1] In these areas there was a strong tendency in the eighteenth century for the hill sheep pastures to give way to arable cultivation on the Norfolk system; but sheep remained important for their manure, and sheep folding on the arable was of course an integral part of the Norfolk system.

The burning of lime was very widely practised, and the Devonshire custom ('denshiring') of stripping off the turf, burning it and spreading the ashes in order to clean and enrich the soil, was also common in western districts, and was the subject of some controversy.[2] Because of costs of transport there was much reliance on local materials as fertilizers. By the seaside, sand from the shore was spread on unyielding clays, and seaweed composted, burnt or ploughed in raw; near centres of textile manufacturing shoddy waste and rags, and near Sheffield the horn and bone waste left over from the making of knife handles, were all put to use; and in every large town the sweepings of the streets and stables and the contents of privies were heaped unhygienically in yards and eventually carried away by barge or cart to the farmers. The farmers of the Surrey clays, indeed, claimed their lands could not be cultivated without liberal doses of the 'London muck'. The great variety of materials made use of, the prevalence of summer fallows on heavy lands even after enclosure, and the attention given to such matters as liming, marling, paring and burning, and other methods of improving the soil, all testify to the serious nature of the problem.[3] Indeed, it was not until the nineteenth century when soil chemistry and artificial fertilizers were established that the treatment of the soil was placed on a more scientific and adequate basis.

So it was also with drainage. Ridge and furrow, the method of surface drainage by which the land was raised in regularly spaced ridges and the rainwater led off in the furrows to the lower ground, was widely used throughout the Midlands and other areas of heavy soils. The wetter the ground the higher the ridges, until in some areas they measured three feet in height from the bottom of the furrow to the top of the ridge; and even if the land was flat and low-lying it was still possible to grow reasonable

---

[1] E. Kerridge, 'The Sheepfold in Wiltshire and the Floating of the Water meadows' *Econ. Hist. Rev.* 2nd ser. VI (1953–4) pp. 282–6.

[2] See W. Marshall, *Rural Economy of the West of England* (1796) I pp. 141–51. The addition of potash to the soil by denshiring was partially offset by a loss of humus.

[3] W. Marshall, *Review of Reports to the Board of Agriculture* (York, 1808–18), esp. V pp. 381–2.

crops on the ridges, while the sheep found in the furrows some protection from the weather.[1] There were some experiments with under-drainage, the most popular form being a deep trench filled loosely with stones, tree boughs, bracken or furze, with the soil replaced above. The Essex and Suffolk drains were two feet deep, wedge-shaped and filled with branches, twisted straw or stones, and in Norfolk Marshall saw pits or soakaways dug at intervals and filled with boughs of ash and alder. It cost about 33s. to drain an acre in the Norfolk manner, and was said to be effective.[2] On the really wet clays, however, these primitive methods were either ineffective or too costly: the problem of heavy land drainage remained unsolved until the introduction of cheap tile drainage in the 1840s, leaving the high-cost and inefficient farming of heavy claylands as the most obvious weakness in the progress of eighteenth-century farming.

There is little precise knowledge of the pre-eighteenth century types of stock or the origins of selective breeding, but its historian makes it clear that there had been for centuries some deliberate crossing of animals in attempts to improve native breeds, and that herds of improved cattle had been established at least as early as the first half of the eighteenth century.

The original types of cattle comprised the hardy black longhorns found in Wales and western coastal districts generally, the red and brown middle-horns and shorthorns of southern, eastern and midland England, together with the dun-coloured dairy cattle of East Anglia and the parti-coloured stock of central England. By the eighteenth century the 'black cattle' driven south and east from Scotland and Wales actually included brown, dun-coloured and parti-coloured beasts, useful both for beef and for milk; and while the red-and-brown cattle of South Wales remained popular for fattening in the English pastures, in North Wales the black runts were being displaced by improved varieties of native sheep that were driven over the border. The red or brown middle-horn type of cattle common in the lowland area from the Wash to the west Midlands and Bristol Channel had probably been crossed with Dutch stock to produce larger beasts that gave more milk. The eastern counties from Lincolnshire northwards were the home of the Yorkshire and Durham shorthorn, a cross between the original red and black cattle and Low Countries stock, animals of great bulk and high milk yield. But although there were

[1] See the discussion by E. Kerridge, 'A Reconsideration of Some Former Husbandry Practices' *Ag. Hist. Rev.* III 1 (1955) pp. 26–32.

[2] Marshall, *Norfolk* II pp. 2–3.

regionally predominant types, the fattening and dairying areas had been stocked indiscriminately and the types were inextricably mixed.

Sheep existed in four basic types: the horned variety of Scotland and the Western Isles; the heath type of middlewool and shortwool animals found in the south-west of England, the south and Midlands; the hairy-fleeced and horned black-faced breeds of northern England, and lastly the longwools of the chalk and limestone uplands of Lincoln, Leicester, Kent and the Cotswolds. By the early eighteenth century the spread of the new crops made for better feeding of the lowland sheep, and had the effect of making possible earlier shearing. But more important, richer fodder also coarsened the fleece. Feeding on clover and the other rich legumes and grasses lengthened the staple of the wool but this meant a coarser fibre, thus giving point to those who complained of the deleterious effects of enclosure on wool. Furthermore, the eighteenth-century improvers concerned themselves more with producing animals good for their meat rather than their wool, and so contributed to the decline in the quality of English wool and the growing demand for imported Irish fleeces and merino fleeces from Spain.[1]

Attempts to increase the size of sheep and the length of the wool staple had been made for four or five hundred years before the eighteenth century, but the growth of scientific selective breeding was very much a development of the early eighteenth century. The famous Robert Bakewell began his livestock experiments about 1745, and achieved his contemporary eminence not because his work was unique but because he was solely a specialist pedigree breeder rather than a grazier-cum-breeder, was successful in popularizing the letting-out of rams and bulls for hire, and not least because he was fortunate enough to attract the enthusiastic notice of publicists such as Arthur Young. But Bakewell certainly had eminent if less well-known predecessors, from whose improved animals he drew the basic stock for his own work: Webster of Canley near Coventry, who about 1750 was making improvements in the Lancashire longhorn, and a certain Joseph Allom who before Bakewell's time had laid the foundations for the improved Leicester sheep. And although in 1790 Marshall acknowledged Bakewell as the leading and best-known breeder, he could name some 15 or 20 men who were doing the same work, and indeed Bakewell's longhorns were surpassed by the

[1] R. Trow-Smith, *A History of British Livestock Husbandry to 1700* (1957) pp. 231–2; *A History of British Livestock Husbandry 1700–1900* (1959) pp. 9, 25–9, 36–40. See also M. L. Ryder, 'The History of Sheep Breeds in Britain' *Ag. Hist. Rev.* XII (1964).

herd of his contemporary Fowler of Rollright. The area around Bakewell's farm at Dishley near Leicester, said Marshall, had 'for many years
abounded with intelligent and spirited breeders'.[1]

Even in the early eighteenth century Defoe was able to note that the
Leicestershire graziers 'are so rich, that they grow gentlemen . . . the
sheep bred in this county and Lincolnshire, which joins to it, are, without
comparison, the largest, and bear not only the greatest weight of flesh
on their bones, but also the greatest fleeces of wool on their backs of any
sheep in England . . .'.[2] But in Defoe's day or a little earlier the beasts still
fattened only very slowly: bullocks would not put on beef until four or
five years old, and wethers could not be finished for the market until at
least four years old. The improvements in the supply and nutritive value
of fodder which the legumes and roots of the new husbandry made
possible helped to speed up the fattening process, and this was taken
further by the practitioners of selective breeding, who gradually produced
animals that combined the various advantages of different breeds in new
varieties, thus making breeding and fattening a more skilled and profitable
branch of agriculture.[3]

As is well known, Bakewell produced a new breed of sheep, probably
beginning with the old Lincoln stock and crossing them with the Ryelands,
and by following the racehorse breeders' principle of breeding in and in to
fix a type that had the desirable characteristics, he produced animals that
fattened rapidly and had a high proportion of saleable flesh to bone. For
the purpose he selected animals with short legs, small heads, and a
rounded barrel shape, and bred only from those, continually weeding out
the progeny that failed to conform—his theory being 'that like produces
like, that small bones, thin pelts and the barrel shape are soonest and most
productive of fat at the least expense of food.' His New Leicesters certainly fattened quickly, and they produced a wool of reasonable quality,
if of lesser quantity; but they were not prolific in breeding, and most
serious of all their meat consisted largely of fat and could be sold only
to the poor, who in any case ate the most mutton and could not afford
to be particular about its leanness. 'A Dishley wether must be slaughtered
at two years old', said Culley, a fellow-breeder, for otherwise it became
'too fat for genteel tables'.[4]

[1] W. Marshall, *Rural Economy of the Midland Counties* (1790) I pp. 270, 295, 381,
392.
[2] Defoe, *op. cit.* II p. 89.
[3] Trow-Smith, *British Livestock Husbandry 1700–1900* pp. 239–40, 244, 255.
[4] *Ibid.* pp. 26–9, 36, 54–64.

67

In addition to his famous sheep Bakewell also played a part in producing the modern shire horse, a breed which sprang from the midland war horse modified by crossing with continental mares. With his cattle, however, Bakewell was less successful. He applied the principles used in breeding the New Leicesters to his New Longhorns, but in the result he produced an animal that on only a moderately rich diet quickly put on masses of fat, but failed to yield the good milk or exhibit the fecundity of the original stock. In the end neither his sheep nor his cattle survived in the form he gave them. The longhorn proved to be a bad choice as the basis of an improved beef animal, and soon the 'longhorn fever' subsided and a new and more fruitful direction was marked by the appearance of improved breeds of shorthorns. His New Leicesters, however, had a great part in the improvement of British and overseas breeds of longwool sheep, particularly the Border Leicester, Wensleydale and improved Lincolns. But even in his own county, Bakewell's New Leicesters soon gave way to Lincoln and Leicester crossbreeds, which had even greater aptitude for early fattening and less of the undesirable characteristics of the Dishley breed.[1]

None the less, Bakewell made his mark as one of the great eighteenth-century farming figures, and his methods served as an inspiration to others, and not only in the improvement of livestock. The Bishop of Llandaff enquired of him his method of watering meadows with irrigation channels, a feature of his farm to which Young had drawn attention, and as his fame spread he received many noble visitors in his Dishley farmhouse. 'The Earl of Hopetown and his Agent, Mr Steward, breakfasted with me on Monday the 16th Instant . . .' he wrote in June 1787, and 'on Wednesday last the Marquis of Graham was at Dishley from 11 to 5 o'clock'. In his letters we find him riding round the Duke of Grafton's estate with the Duke and Arthur Young, meeting Sir John Parnell at Buxton to inspect some rams, and taking the Earl of Hopetown to see the stock in Smithfield market 'where his Lordship took great pains to inform himself of the utility of the different kinds he there met with, and this I think the best school for any Gentleman who has a liking to this kind of amusement.'[2]

As Bakewell set his mark upon the longwool sheep, so a little later, round about 1780, John Ellman of Glynde in Sussex refashioned the shortwool breeds. Ellman and his successors (including Jonas Webb) eventually transformed the Southdown from a light and long-legged

---

[1] *Ibid.* pp. 46–69; Trow-Smith, *English Husbandry* pp. 156, 164–6, 170.
[2] H. C. Pawson, *Robert Bakewell* (1957) pp. 113, 134, 156–8, 172.

animal into one solid and compact, excellent for mutton and still very good for wool, and thus began the process of development which gave us new breeds of downland shortwool sheep. The selective inbreeding of cattle which Bakewell and numerous others had tried was brought to success after 1780 by the Colling brothers, who farmed near Darlington, the Booths and Thomas Bates. Working on the shorthorn variety, which soon quite displaced the longhorns favoured by Bakewell, they managed to produce animals good for both meat and milk, and laid the foundations of an important industry in pedigree stock. Charles Colling was the first breeder to receive as much as 100 guineas for a shorthorn cow, and on his retirement from breeding his celebrated bull Comet was sold for 1,000 guineas. By 1800 the importance of quality in livestock was becoming fully recognized by progressive farmers, and animal husbandry had made a giant stride away from the haphazard breeding and indiscriminate standards of former times.[1]

In the eighteenth century the farmer's tools were made of wood and wrought iron, and were generally crude and inefficient. The style and size of the implements varied greatly from area to area, and the local designs often persisted long into the nineteenth century, and even in some cases into the twentieth century. Indeed, a remarkable feature of the agricultural revolution was the slow pace at which improved tools and machinery were brought into use (as compared, for instance, with the comparative rapidity of the enclosure movement or the introduction of improved livestock). A wooden plough could be seen at work near Brighton even in the 1850s, and at the same date corn was still widely threshed by hand. Not only conservatism but the small size of many farms, the varied conditions to be found in English farming, and the cheapness of agricultural labour were factors in the pace of this kind of change.

Progress was tied up with the developments of the Industrial Revolution, the substitution of cast iron and the standardized factory product for the implements of wood, wrought iron or stone, fashioned individually by the village blacksmith and carpenter. It was not until the close of the eighteenth century that the iron plough was widely replacing wooden ones, and by this time the Rotherham plough, based on Dutch designs and patented in England as early as 1730, had come into favour in the north

[1] Trow-Smith, *English Husbandry* pp. 156–7, 165–7; *Livestock Husbandry 1700–1900* pp. 127–30, 236–41, 276–7. For a discussion of the later developments in breeding, see C. S. Orwin and E. H. Whetham, *History of British Agriculture, 1846–1914* (1964) pp. 12–15.

and east of the country. The Rotherham was a swing plough with a curved mouldboard which performed its function of turning over the earth much better than did the straight board formerly used; but its chief advantage lay in the design of its main frame, making it smaller and lighter, and requiring fewer draught animals than ploughs of traditional construction. The period was one of considerable interest in new designs of ploughs, and among the numerous books on the subject was the important *Treatise on Ploughs and Wheeled Carriages* (1784) by James Small, who himself followed up the experiments of the Norfolk farmer Arbuthnot to determine the optimum curve of the mouldboard. It was Small 'who reduced to paper the natural curve of mouldboard which he discovered by allowing the turning furrow slice to scour a soft wooden mouldboard until it arrived at the lines which still obtain today.'[1]

In the 1780s Robert Ransome, founder of the Ipswich firm which still bears his name, introduced his self-sharpening hardened cast-iron plough-shares, a great step forward. In 1808 ploughs with standardized parts for easy replacement were available, and by 1840 the Ransome factory was making as many as 86 distinct types of plough to suit the needs of local markets. Rollers, formerly made of stone or consisting merely of a log drawn by a horse, and harrows too, often nothing more than bushes tied to a wooden frame, were now available in cast iron.

Seed-drills and horse-drawn hoes, developed from Tull's early designs, were produced commercially in the early nineteenth century, but were rarely found outside the light soil areas, and only infrequently even there. A hundred years after Tull first invented his seed-drill Arthur Young could find only a dozen farmers in the whole of the agriculturally progressive county of Hertfordshire who drilled their crops.[2] Tull's advocacy of drilling seeds in widely-spaced rows, and regular hoeing of the soil between the rows, had the two great advantages of saving expensive seed and of keeping down weeds, thus reducing the need for fallows. But his advice was suitable only for well-drained and easily cultivated soils, and even there the vast majority of farmers preferred to sow their seed broadcast and put up with the weeds rather than face the practical difficulties and labour costs of drilling and hoeing. In his *Rural Economy* (1770), Young included a detailed discussion of the reasons for the failure of the drill to replace hand sowing. The machines, he pointed out, were costly and difficult to obtain, and their construction was not sufficiently robust to withstand the hard usage of clumsy labourers unaccustomed to hand-

[1] Trow-Smith, *English Husbandry* p. 138.
[2] *Ibid.* p. 135.

ling expensive equipment; the machines, it appeared, could not be designed to allow for different types of soil and the various sizes of seeds which might need also to be sown at varying depths and intervals; most important of all, for drills to be used on heavy clays and loams the land required three spring ploughings and harrowings, and this made for very late sowing and high cultivation costs.[1]

Mowing and reaping were among the hardest and most tedious tasks of the farm and the most labour-consuming, making the hay-time and harvest periods when all the labourers, the farmer's family, and any other available hands from among the local craftsmen and industrial workers were all called in to help. Harvest hours were long, beginning as soon as it was light and going on until it was too dark to see, but there were the compensations of good wages and plenty of food and drink. But after harvesting there still remained the long-drawn-out work of threshing and winnowing in the barn.

The whole process was incredibly arduous and slow:

Corn was cut by scythes, fagging hooks, or sickles; if with the first, each scytheman was followed by a gatherer and a binder; a stooker and raker completed the party. When a good man headed the gang, with four men to each scytheman, two acres a day per scythe were easily completed. Threshed by the flail, the grain was heaped into a head on the floor of the barn. The chaff was blown away by means of the draught of wind created by a revolving wheel, with sacks nailed to its arms, which was turned by hand. Thus winnowed, the grain was shovelled, in small quantities at a time, into a hopper, whence it ran, in a thin stream, down a screen or riddle. As the stream descended, the smaller seeds were separated and removed. The wheat was then piled at one end of the barn, and 'thrown' in the air with a casting shovel to the other extremity. The heavy grain went furthest: the lighter, or 'tail', dropped short. To some of the corn in both heaps the chaff still adhered. These 'whiteheads' were removed by fanning in a large basket tray, pressed to the body of the fanner, who tossed the grain in the air, at the same time lowering the outer edge of the tray. By this process the whiteheads were brought to the top and extremity of the fan, whence they were swept by the hand. Lastly the corn was measured, and poured into four-bushel sacks, ready for market. The operation of dressing was slow. As the sun streamed through a crack in the barn-door, it reached the notches which were cut in the wood-work to mark the passage of time and the recurrence of the hours for lunch and dinner. The operation was expensive as well as slow, costing from six to seven shillings a quarter. Hay was similarly made

[1] A. Young, *Rural Economy* (Dublin, 1770) pp. 167–77.

in all its stages by hand, and with a care which preserved its colour and scent. The grass, mown by the scythe, fell into swathes. These were broken up by the haymakers, drawn with the hand-rake into windrows, first single, then double. The double windrows were pulled over once, put first into small cocks, then into larger which were topped up and trimmed so as to be shower proof, and finally arranged in cart-rows for pitching and loading. Women, working behind the carts, allowed scarcely a blade to escape their rakes.[1]

Successful reaping machines were not produced until the early nineteenth century, and the 1812 design of John Common, a Northumberland millwright, became the basis for the famous McCormick reaper, which after competing well with other American reapers on the prairies was brought over to England and shown at the Great Exhibition of 1851, achieving wide acclaim.[2] Only then, in the middle nineteenth century, with 'high farming' coming into vogue, was the earlier machine of the Rev. Patrick Bell, invented in 1826 and improved as the 'Beverley Reaper' in 1853, fully appreciated. However, the threshing machine of another Scotsman, Andrew Meikle, obtained wide popularity soon after its appearance in 1786. Improving on earlier designs, it could be driven by steam, by water, horses or by hand, and together with the later winnowing machines greatly shortened the time-honoured process of threshing the corn with flails and winnowing it by throwing the grain into the air by hand. Indeed the widespread adoption of threshing machines in the corn areas of the south of England deprived many labourers of their winter standby, and was a factor in the revolt of 1830. There were many other devices useful for preparing cattle food, chaff cutters, root slicers and crushers, and these were first turned by hand and eventually by steam on the large farms.

Finally, the eighteenth century saw considerable advances in means of farm transport, in addition of course to the improvement of rivers and roads and the building of canals. Although wheel-less sledges and barrows remained indispensable in the steep hilly areas of the west and north until the nineteenth century, elsewhere the two-wheel carts were supplemented or replaced by large four-wheel wagons which could carry much heavier loads. The style of wagons varied greatly from one district to another, and in particular the nature of the roads locally determined the breadth of the wheels—broad wheels up to eight inches in breadth being popular in the soft mud of the clay areas, and narrow wheels on the harder roads of the drier soils.

[1] Ernle, *op. cit.* pp. 360-1.
[2] Trow-Smith, *op. cit.* p. 205.

Thus the farmers gradually improved their tools, transport and mechanical devices along with their improved rotations and livestock. But the changes in this sphere were supplementary and secondary to the changes in the fields; they were aids to efficiency and higher production rather than prime causes of agricultural change, and until well into the nineteenth century it could not be said that the agricultural revolution depended to any significant extent on machinery.[1]

Lastly we must say something of the work of the eighteenth-century writers and publicists. Although there were some important and fairly influential writers in the earlier decades of the eighteenth century— William Ellis, John and Edward Laurence, Edward Lisle, and Tull himself, to mention only a few—the great names occur in the years after 1760. It is now generally agreed that the soundest writer and the one with the most comprehensive understanding of agricultural practice was William Marshall. His systematic and detailed accounts of farming methods in the Midlands, the south, west, Norfolk, Yorkshire, and other areas made clear the best methods of the progressive farmers and castigated those that were backward and inefficient. In particular, Marshall had a keen sense of the relevance of natural farming areas as determined by soil, climate and relief, and he rightly attacked the Board of Agriculture (formed with government support under the chairmanship of Sir John Sinclair in 1793) for drawing up its *Reports* on a county by county basis. County boundaries did not represent farming boundaries, Marshall argued, and in his monumental *Review of the County Reports* he constantly stressed this general failing, as well as the incompetence of many of the reporters and their more parochial weaknesses.

Some of the county *Reports*, however, were written by acknowledged experts who had a real knowledge of the farming of their counties. A number of them, like Thomas Davis who reported on Wiltshire, were stewards to large landowners, as both Young and Marshall had been at various times. Other competent *Reports* were written by leading farmers, such as Boys (Kent), Bailey and Culley (Northumberland), and Mavor (Berkshire).

Another of the leading authorities was Nathaniel Kent, whose *Hints to Gentlemen of Landed Property* (1775) was widely read, and he set up what must have been one of the earliest firms of specialist London estate agents, who for a standard scale of fees supervised estates for owners. The

[1] For an account of subsequent developments in farm machinery see Orwin and Whetham, *op. cit.* pp. 102–14.

most colourful, most readable, and certainly the most-quoted writer, however, was Arthur Young, whose varied career embraced the roles of pamphleteer and author, Parliamentary reporter, farmer, estate steward, and Secretary to the Board of Agriculture. Although there is some justice in the oft-repeated jibe that Young failed as a practical farmer, it is but a half-truth, for his failures came when he was very young and inexperienced, and when at a later period he inherited the family property in Suffolk he seems to have managed it with no difficulty and with as much competence as his writing and frequent absences from home would allow.

In addition to his early manuals and *Tours*—the *Farmer's Letters*, *Rural Economy*, *Farmer's Guide in Hiring and Stocking Farms*, *Southern*, *Northern* and *Eastern Tours* were all written between 1767 and 1771 before he was thirty—Young wrote vastly interesting accounts of Ireland as it was about 1780 and of France on the eve of the Revolution, a general work on *Political Economy*, more specialized works on farming questions, and several of the Board of Agriculture's *Reports*. Consistent only in his support for agricultural improvement of every kind, Young wrote much that was biased, inaccurate and contradictory. His method of collecting material was haphazard and his work needs to be read with some reserve. In his *Tours* and county *Reports* it is often difficult to disentangle from his accounts of progressive experiments and instances of backwardness the really typical practice of an area, and he was apt to let his own enthusiasms run away with him. On such questions as leases, tithes, enclosures and open fields he took extreme positions which he was often obliged to modify later, and he ignored, as Marshall did not, the long history of empirical experimentation on which such 'new' developments as Bakewell's improved livestock were founded.

However, he inevitably became identified with the progress of agriculture, a figure of international reputation, and the spokesman of the country gentlemen and large farmers in their campaigns for the cheapening of enclosure, the commutation of tithes, and the removal of the prohibition on wool exports. His voluminous *Annals of Agriculture* provided a forum for progressive views and for news of experiments and innovations. Young is not the most reliable guide to the development of farming in the later eighteenth century, but in his unbounded enthusiasm and ceaseless enquiry he truly represents the pioneering spirit of change. And for all the inconsistency and exaggeration there is still enormous value and interest in his work.

How far the writings of Young, Marshall, Kent, Tull and a host of others contributed to the progress of the time it is impossible to say. Their direct influence on the adoption of improved methods was probably very limited. According to Young himself the *Annals of Agriculture* had a circulation of only about 400, and it seems likely that much of the preaching in his and other writers' work was to the already converted. Indeed, it may well have been through personal contacts, through meeting the great landlords and riding with them over their estates, through talks with gentlemen in their homes over dinner and with farmers met with on the road, that Young and the others achieved most. Personal discussion against the background of local conditions and local experience was likely to be far more fruitful (as it still is today) than printed accounts of remote and rather unreal experiments. François de la Rochefoucauld, a Frenchman visiting England in 1784, has left a vivid account of the enthusiasm which Young could inspire and which was perhaps his main contribution to improvement. With Young he made a short tour through Suffolk to see the nature of the farming, and he wrote: 'It is incredible how intelligent the farmers are, even the small farmers . . . I have seen a hundred of them talking with Mr Young on the principles of their calling, making suggestions and recounting their experiences for three-quarters of an hour or an hour; they never failed to win Mr Young's admiration, though he was well used to it.'

*Suggestions for further reading:*

G. E. Fussell            'From the Restoration to Waterloo', Introduction, Lord Ernle, *English Farming Past and Present* (6th ed. 1961).

G. E. Fussell            'Low Countries' Influence on English Farming' *English Hist. Review* LXXIV (1959).

G. E. Fussell            *Jethro Tull: His Influence on Mechanized Agriculture* (1973).

G. E. Fussell            'Science and Practice in Eighteenth-Century British Agriculture' *Agricultural History* XLIII, 1 (1969).

E. L. Jones              'English Farming before and during the Nineteenth Century' *Economic History Review* 2nd ser. XV (1962–3).

| E. Kerridge | 'Turnip Husbandry in High Suffolk' *Economic History Review* 2nd ser. VIII (1955-6). |
| E. Kerridge | 'The Sheepfold in Wiltshire and the Floating of the Water-Meadows' *Economic History Review* 2nd ser. VI (1953-4). |
| E. Kerridge | *The Agricultural Revolution* (1967), Ch. 3-9. |
| W. Harwood Long | 'The Development of Mechanization in English Farming' *Agricultural History Review* XI (1963). |
| G. E. Mingay | 'The Eighteenth-Century Land Steward' in *Land, Labour and Population in the Industrial Revolution* (ed. E. L. Jones and G. E. Mingay, 1967). |
| C. S. Orwin and E. H. Whetham | *History of British Agriculture 1846-1914* (1964), Ch. 1, 4. |
| R. A. C. Parker | 'Coke of Norfolk and the Agricultural Revolution' *Economic History Review* 2nd ser. VIII (1955-6). |
| B. H. Slicher van Bath | 'The Rise of Intensive Husbandry in the Low Countries' in *Britain and the Netherlands* (ed. J. S. Bromley and E. H. Kossman, 1960). |
| Joan Thirsk | *The Agrarian History of England and Wales* IV (ed. Thirsk), Ch. 1, 3. |
| R. Trow-Smith | *English Husbandry* (1951), Ch. VIII-IX. |
| R. Trow-Smith | *A History of British Livestock Husbandry 1700-1900* (1959), Chs. 1-3, 8-9. |
| *History of Technology* IV (ed. C. Singer, E. J. Holmyard, A. R. Hall, and T. I. Williams, 1958), Ch. I-II. |

# 4 Enclosure

The enclosure of open fields, common lands, meadows and wastes constitutes one of the most remarkable developments in English agriculture in the eighteenth and early nineteenth centuries, and certainly the one that has aroused the most controversy. It affected a very large area of land—well over 6,000,000 acres or something like a quarter of the cultivated acreage—and it left its mark on the nature of English farming, and indeed on the very appearance of the countryside. And although the movement continued over the whole of the eighteenth century and much of the nineteenth century it was concentrated heavily in certain periods, notably the 1760s and the 1770s, and in the Napoleonic war years between 1793 and 1815. In the first of these periods some 900, and in the second period some 2,000 Acts of Enclosure were promoted, quite apart from the unknown but considerable number of enclosures arranged by agreement between proprietors without an Act. There were over 4,000 Acts in total between 1750 and 1850, but nearly three-quarters of them were concentrated in the two periods of heavy enclosure, which made up together only 39 years of the whole 100. It is evident, therefore, that the impact on the countryside must have been particularly felt in these two major outbursts of enclosure activity.

Nevertheless, we must not lose sight of the fact that enclosure was a continuous process which had been going on with variations in pace and emphasis from the Middle Ages. By the beginning of the eighteenth century the area where there were still numerous open fields and commons consisted largely of the midland and central southern counties of England, and it has been estimated that about half of the existing arable land still lay in open fields. The area concerned was bounded by Yorkshire on the north, and a line drawn from the Lancashire-Yorkshire border down to Bristol on the west; on the east only the counties of Kent, Surrey,

Sussex, Hertfordshire, Essex and Suffolk had little or no open fields left to enclose, and on the south there was little open field left in Hampshire and Dorset also. The enclosure of commons and waste alone, as distinct from open fields, was especially marked in the northernmost counties of England and Wales, in Yorkshire, Derbyshire, Lincolnshire and Norfolk, some parts of the west Midlands and Somerset. Elsewhere, apart from a few isolated districts, there were no large areas of common land or waste available that were worth enclosing.

The type of enclosure which most often gave rise to controversy and complaint was that concerned with open fields and commons, and thus was largely confined to the central midland area of the country. Throughout our period, but particularly in the first half of the eighteenth century, when Acts of Parliament were little used, enclosure was proceeding 'by agreement', i.e. by private arrangement between the owners of the land concerned. Where all the land in a village was in the hands of one or two owners it was generally easy to arrange an enclosure of this sort, but the word 'agreement' was sometimes stretched to cover arrangements arrived at only as the result of pressure on some of the parties, including the threat of lawsuits. The resulting agreement was frequently enrolled in Chancery to give it greater legal force. Even where there was a multiplicity of owners there were small-scale enclosures going on, the owners of neighbouring lands arranging exchanges for mutual convenience and taking land out of the open fields to provide additional closes. Thus without any formal Act of Parliament or any sweeping change, the open fields of numerous villages gradually diminished. It is impossible to say how much land was enclosed by agreement rather than by Act, but it must have been very large, perhaps half as great as the open-field area enclosed by Act.

Landowners increasingly resorted to the Act of Parliament in the second half of the eighteenth century because it offered several advantages over enclosure by agreement and had only one serious disadvantage—higher cost. Enclosure by Act made it possible to enclose the whole of the open fields, together with the commons and any suitable waste land, at one and the same time, and thus facilitated an important and rapid improvement in farming efficiency. The Act also conferred legal sanctity and finality on the new arrangements, and lastly it made it possible for the owners of the greater area of land to force the hand of other proprietors in the village who might be opposed to the change. This last feature has been the basis of the criticism of those writers who have seen the Enclosure Act as an instrument of oppression, by which the small freeholder was coerced into participating in changes which were against his interests, and in which the cottager suffered by the loss of access to commons and waste. How far the Enclosure Acts did affect villagers adversely we shall discuss below, but first we must examine the purposes of enclosure and the reasons for the great volume of enclosure in certain periods: between 1764 and 1780, and between 1793 and 1815.

There were broadly four main objects behind an enclosure, whether by Act or otherwise. First, enclosure undoubtedly made for more efficient farming by making farms more compact, larger and easier to work, by making possible a better balance between arable and pasture, by encouraging the adoption of alternate or convertible husbandry, and by allowing better care of animals—in short by overcoming the defects of much of the open-field farming. This is not to say that enclosure always produced all these results, far from it. There is a great deal of evidence that where the soil was unsuited to roots and the new legumes, or where the farmers were backward and conservative, then enclosure might make little difference to farming methods or output. In some areas the soils were so intermingled that it was difficult to create workable farms; and where the cost of fencing the allotments was high, it was not uncommon for the new farms to remain for many years without fences or hedges. This of course made the land difficult to manage and liable to the depredations of roaming livestock, and meant a failure to realize some of the principal advantages of enclosure. But generally enclosed farms were more efficient, easier to manage and more productive, and these advantages were reflected in the higher rents which farmers were prepared to pay for them. Even when the rotation remained exactly the same as in the neighbouring open fields the occupiers of enclosed farms were willing to pay a considerable premium over the rent of the same land when occupied in intermingled holdings and subject to common rights.

A second object of enclosure was to convert land to more profitable uses. Thus old arable land, which through a rigid system of cropping had become worn out, might be rendered profitable again if laid down to permanent grass or long leys. Commons and rough hill grazing, overgrown with weeds or bare through continual over-grazing, might be improved by ploughing up and putting under a suitable rotation; and waste could be cleared, ploughed up, and made valuable by a judicious application of marl and manure. The enclosure of commons and wastes achieved a third object, namely that of expanding the area of land under regular cultivation, the unproductive or lightly-cultivated areas, such as the chalk and limestone downs and wolds, being converted into permanently useful acres. Finally, little-noticed objects of some enclosures were to improve efficiency by getting rid of tithes and by bringing order and greater convenience into a parish, where through the years the gradual, piecemeal reduction of the open fields had resulted in an unwieldy complex of tiny closes.

Enclosure Acts, therefore, attempted to achieve some or all of these objectives. Of the more than 4,000 Acts about two-thirds were concerned with both open fields and commons and waste, while the remainder concerned commons and waste only. Many factors influenced the chronology of enclosure but, generally speaking, villages which had good communications, a small number of owners, and a good or readily improvable soil tended to be enclosed first.

The question of the common was also important: the existence of large areas of fertile common land might be the main inducement to enclose; on the other hand the handicap of inadequate, overstocked or severely stinted commons might hurry on enclosure to remedy the shortage of pasture. To give an example, one factor which influenced the enclosure of the village of Flintham in Nottinghamshire was the presence of so large an area as '450 Acres of Common, which is now of little advantage to the larger Farmers, because it is unstinted and allways over stock'd by the Cottagers, the soil of which is sound and dry, fit for marling and turniping . . .'.[1] Soil was crucial in the eighteenth century. 'It is well known', stated Thomas Stone, a contemporary expert, 'that the most profitable soil to inclose is a sandy or light loam, where the cultivation of turnips, clover and the artificial grasses can be effected with certain success. . . . On such soil abundance of food for stock is produced, which generally furnishes a great quantity of rich manure for the arable land, so that the business of farming and grazing together are much more advantageously pursued than either can be to any great extent without the other.'[2] While the date, purposes and effects of a particular enclosure were always subject to a variety of local influences, there is no doubt that purely geographical factors such as soil, climate and relief were of basic importance. This was first pointed out by Professor Gonner, who attempted to relate soil characteristics to the period and effects of enclosure. He found that the midland claylands were generally enclosed early in the period of parliamentary enclosure, often with the object of converting arable to pasture, while the chalk and limestone uplands, originally sheepwalks, were

[1] Notts. R. O.: DDTN 4/9–11 (1759).
[2] T. Stone, *Suggestions for Rendering the Inclosure of Common Fields and Waste Lands a Source of Population and Riches* (1787) pp. 26–7.

more commonly enclosed for creating additional arable land, particularly after 1760.[1]

When studied in detail for a particular region, however, it becomes clear that these broad conclusions often break down, the geographical influence mingling with other factors, and no simple or direct connection necessarily existed between the physical characteristics of an area and the date and effects of enclosure. In Lincolnshire, for example, Dr Thirsk found that the chalk and limestone uplands were subject to much early enclosure by Act of Parliament, i.e. between 1750 and 1780, at about the same time that a good deal of the fenlands and claylands of the county were enclosed. The marshland area, a fourth geographical division, did not, however, see much parliamentary enclosure until after 1815. In Lincolnshire the differences in the periods and extent of enclosure arose only in part from physical characteristics and the existing nature of the farming and possibilities of improvement; the presence of landowners of enterprise, and the numbers of small freeholders and total size of the population had often a more significant influence. The upland sheepwalks offered much scope for conversion to profitable farmland; they were thinly populated, the farmers were enterprising, and so they were rapidly enclosed; the marsh areas, on the other hand, had only limited areas of open field and common to enclose, and were heavily populated by poverty-stricken small freeholders who lacked the means and scope for improvement.[2]

It is evident, then, that since in each village the nature of the soil, the extent of the commons and the availability of waste, the area of remaining open fields, the size of the existing farms and the standards of cultivation, the numbers, wealth and enterprise of the freeholders, all differed to a greater or lesser degree—that the objects of enclosure, and its effects on the farming and on the village community, would differ also. Each village and each enclosure had some degree of uniqueness. This makes it very difficult to generalize about the consequences of the change, as will be evident in later sections of this chapter. And it is clear also that one cannot say either that all enclosures had the same purposes. The majority of enclosure Acts contrived the replacement of open fields and commons by compact farms; others were concerned entirely or mainly with extending the cultivated area on the hills and moors; others dealt with a small rump of open fields and the commons; yet others attempted to improve on the existing layout of farms; and finally there were some

[1] E. C. K. Gonner, *Common Land and Inclosure* (2nd ed. 1965) pp. 236-7.
[2] Joan Thirsk, *English Peasant Farming* (1957) pp. 237-8, 292-3.

which merely confirmed an enclosure carried out at some previous date by agreement. The complexity of the question and the important geographical variations make enclosure one of the most intractable and yet fascinating of historical problems.

The broad development of enclosure can be traced in the number of Bills brought before Parliament in each year. Of course, this is only a rough guide to the extent of land enclosed since the area concerned varied with each Bill and the figures do not take account of the considerable areas enclosed by agreement. There is also the question of time-lags. The introduction of a Bill into Parliament, coming as it often did at the end of several years of negotiation and of buying-out of freeholders, might represent the final stage of the enclosure process rather than the beginning, and the expenditure on the work of enclosing might be spread over a lengthy period of years after the obtaining of the Act.

Over a period, however, the number of Bills will be sufficient to provide a rough indication of the prevailing interest and investment in enclosure. On this basis Professor Ashton has pointed out that a relationship exists between the number of Bills and the movements in the rate of interest on the Funds, i.e. the stocks issued by the government (Consols), the Bank of England, and the East India and other companies.[1] The implication of this relationship is that the cost, or more realistically the difficulty or ease of borrowing money, together with the level of agricultural prices, had an influence on the volume of enclosure undertaken at a particular time, difficult conditions for borrowing (as in the 1780s) having the effect of discouraging enclosure, and easy borrowing conditions (as in the later 1760s and early 1770s) encouraging it (see fig. 1).

There was undoubtedly some significance in this relationship but it should not be pressed too far. It depends on two assumptions: that investment in enclosure was closely linked with landowners' ability to borrow, and that the return on investment in enclosure was comparable to the return on the Funds. Now, it is possible that much enclosure was financed by borrowing on mortgage, and it is true that borrowing of this kind was restricted or even impossible when the high yield on the Funds made it the more attractive investment for lenders. But it seems probable (and more research is needed on this point) that a large proportion of enclosure, especially that promoted by large landowners, was financed not by borrowing but out of current estate income, perhaps out of the

[1] T. S. Ashton, *An Economic History of England: the Eighteenth Century* (1955) pp. 40–1.

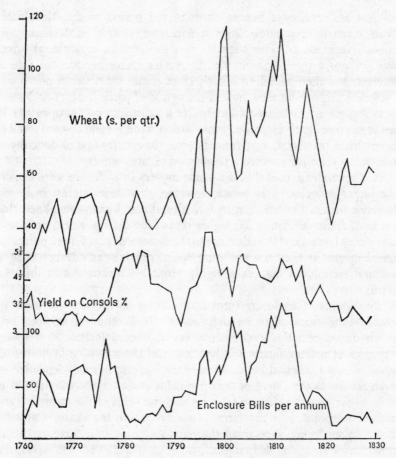

**I** *Annual numbers of Enclosure Bills, yield on Consols and wheat prices 1760–1830*

rents of the property concerned, and this is especially likely as the expenditure was in fact spread over a number of years. Moreover, the willingness of landowners to devote more of their estate income to investment in enclosure was encouraged by the high yield obtained, the return being much higher than those on investments in the Funds, in land purchase or other forms of estate development, or virtually any of the range of investments open to landowners (with the possible exception of some canal investments).

Finally, it should be noticed that the relationship between the rate of

83

interest and enclosure breaks down in the period of the Napoleonic Wars when interest rates rose but enclosure, instead of declining, increased enormously. This suggests that the level of agricultural prices was perhaps a more significant influence on enclosure than the rate of interest, and there is indeed a fairly close alignment between prices and enclosure throughout the whole of the period of parliamentary enclosure, upswings in prices being followed after a short interval by upswings in enclosure (see fig. 1). However, while no doubt the general upward trend in prices had a persistent, long-run influence, the relative ease or difficulty of borrowing certainly affected a proportion of landowners.

It is generally agreed that the prime movers in enclosure were usually the larger proprietors. It seems, however, that their interest in it was largely a financial rather than an agricultural one. Landowners knew that enclosed farms yielded much higher rents than did open-field ones, and while they recognized the relevance to the increased rent of more profitable farming, usually they saw enclosure not so much as an advance in agricultural techniques than as a highly profitable investment in financial terms.

The profits of enclosure turned, of course, on the ratio between its cost and the increased rent which resulted. It therefore varied according to circumstances: the level of cost was influenced mainly by the local expenses of making fences and hedging, and the provision in numerous cases of new roads and buildings; and the increase in rents depended on such factors as the extent of the open fields enclosed, the adaptability of the soil for more productive farming, and the value of the commons and waste incorporated in the farms. In areas where the changes brought about by enclosure were small the rents rose only to a small extent; in other cases they doubled, trebled, or even quadrupled. On average it may be estimated that rents perhaps about doubled; the landowner's gross return on his investment was probably between 15 and 20 per cent, but higher where much waste was enclosed. This was an extremely good return; very much higher than on an investment in the Funds or in land purchase (the gross yield of which was only 5 or 6 per cent). Enclosure was thus by far the most profitable use of capital in connection with land, and perhaps more profitable than many riskier commercial or industrial ventures, and this goes far to explain its popularity in the generally thriving conditions for agriculture between 1760 and 1813.[1]

In actual figures the cost of enclosure might vary from as low as a few

[1] See the discussion of enclosure costs and returns in F. M. L. Thompson, *English Landed Society in the Nineteenth Century* (1963) pp. 222-6.

shillings to several pounds per acre. In the late eighteenth century the average cost of an enclosure by Act of Parliament was estimated by the Board of Agriculture at about 28s. per acre. This figure included approximately 8s. 6d. per acre for obtaining the Act, 10s. for the fees of the commissioners and the expenses of surveying and valuing, and 9s. 6d. for the cost of fencing.[1] Where the enclosure involved new farmhouses and roads, however, the total outlay could be easily three or four times as much. It is not surprising, therefore, that in many enclosures there was no attempt to rebuild farmhouses or lay down new roads, and that as a consequence the new farms were often remote from the farmers' dwellings, yards and barns. For this reason, where the farmhouses remained in the centre of the village it was sometimes the practice to give the small farmers allotments near at hand, and to allot the more distant fields to the wealthier owners who could afford to rebuild. In general, the rent of open-field land varied from about 5s. to 10s. per acre in the later eighteenth century; after enclosure from about 10s. to 20s. or more. Perhaps a doubling of rents, from about 7s. to 15s. per acre, was the common result of enclosure in the Midlands.

The procedure of enclosure under Act of Parliament was one of a petition to present a Bill, the granting of the Bill and the appointment of commissioners, followed by the actual re-allotment of the land under the commissioners' supervision. In their eloquent and detailed account of the changes in the village in this period, J. L. and Barbara Hammond stressed the opportunities which existed at each stage of the procedure for the wealthy owners to gain their ends against the wishes and interests of the majority of the village inhabitants. They pointed out that before 1774 it was not even necessary for public notice of a petition to be given; that in Parliament the Bill might be considered by a committee composed entirely of the large proprietors interested in it, and that petitions against the Bill presented by small freeholders and others could be ignored; that the Act itself often gave undue preference and privileges to the lord of the manor and the titheholder; that the commissioners appointed to carry out the enclosure might be nominated solely by these parties, and thus showed them undue preference in the award of allotments; and finally that the claims of the small farmers and cottagers were likely to fall foul

---

[1] W. E. Tate, in 'The Cost of Parliamentary Enclosure in England (with special reference to the County of Oxford)' *Econ. Hist. Rev.* 2nd ser. V (1952–3), found 25s. to be the average cost of 38 enclosures in Oxfordshire between 1757 and 1796, but to this figure should be added 2s. 6d. (10%) to allow for the costs of the incumbent which were normally borne by the other proprietors.

of the legal niceties required of the claimants, and so be rejected or over-looked.[1]

Quoting from a number of Acts and Awards the Hammonds certainly convey the impression that the system of parliamentary enclosure was in fact a gigantic swindle by which the wealthy owners gained land and wealth, and in which they rode roughshod over the rights of small men. Other writers who have carried out very extensive examinations of many hundreds of enclosures have arrived, however, at very different conclusions. Professor Gonner, whose work on enclosure was outstanding for its care and detail, was impressed by the complexity of the task of enclosure and the fairness with which the commissioners generally carried it out. The commissioners had to take into account that some land possessed more common rights than did other land, and that some land had no common rights attached to it at all; that some unenclosed land was worth more than other, and in estimating values that small farmers were rented more highly than large ones. The commissioners had often to assess what quantity of land was a fair allotment in lieu of tithes; they had to put land aside for roads, and perhaps for a quarry or gravel pit to keep the roads in repair; and sometimes they reserved some land for the poor as cow pastures and vegetable gardens, and for the parish for a school or a poorhouse. 'When the gravity and delicacy of the task undertaken by the commissioners is considered', wrote Gonner, 'the existence of complaint against them is not astonishing. It is rather a matter for wonder that the complaints were not far louder and universal.'

The conduct of an enclosure was such a complex matter that in practice it became a professional occupation for the country gentlemen, land agents and large farmers who were experienced in it, and we find the same commissioners acting at a variety of different places. Gonner found that the commissioners did not take a strict view of the legal validity of claims but often gave favourable consideration to claims for compensation based on custom or equity rather than on legal right. Even the Hammonds conceded that the squatters who had settled on the waste and were suffered in the village as 'poor aliens' were sometimes treated fairly, some with more than 20 years standing as occupiers being allowed to keep their encroachments, and those of a lesser standing being allowed to purchase them. 'Taken as a whole', said Gonner, 'the work of division and apportionment appears to have been discharged conscientiously and fairly.'[2]

A more recent investigator, Mr W. E. Tate, has endorsed Gonner's

[1] See J. L. and B. Hammond, *The Village Labourer* (1912), Ch. II–IV.
[2] E. C. K. Gonner, *Common Land and Inclosure* (2nd ed. 1965) pp. 82, 95.

conclusion. A remarkable feature of eighteenth-century enclosure, he has said, was the 'care with which it was carried out and the relatively small volume of organised protest which it aroused.' On the basis of his evidence he found that the instances of enclosures deliberately rigged against the small man, which the Hammonds quoted, were 'in the highest degree exceptional . . .', and that it would be quite unfair to suppose them typical of the country in general. Ultimately, of course, the enclosure commissioners relied for their employment on the large proprietors, and were bound to satisfy them. But apparently this was not incompatible with the striking of a fair balance between the claims of both large and small men, and it is clear that small owners did not always fear the outcome. Together with other writers, Mr Tate finds that it is untrue that small proprietors were invariably opposed to enclosure, that they refused to sign petitions or were afraid to oppose them. 'The weight of propertied opinion was overwhelmingly in favour of enclosure in whatever units that weight was expressed.'[1]

The controversy continues. In a recent work Dr Hoskins, a distinguished authority on Leicestershire, pointed out that at Wigston Magna the extinction of the tithes at the enclosure involved an appropriation of 380 acres, or one-seventh of the available land, and that the titheholder's allotment was fenced at the expense of the other proprietors, a very common provision. As a consequence, the small farmers at Wigston came out of the business with a smaller acreage than before (although of course with land free of tithes), and some of their holdings were too small to be successfully worked. In a heavily populated and densely settled county like Leicestershire, where available waste land was small or non-existent, Dr Hoskins comments, enclosure and engrossing were bound to have serious effects on the small farmers.[2]

Other investigators have tended to support Mr Tate by emphasizing that it was sometimes the larger owners and the parsons (concerned about their allotment in lieu of tithes), rather than the small owners, who opposed enclosure. This was the case, for example, in the East Riding;[3]

[1] W. E. Tate, 'Opposition to Parliamentary Enclosure in Eighteenth-Century England' *Agricultural History* XIX (1948) pp. 137, 141–2. See also his 'Parliamentary Counter Petitions during Enclosures of the Eighteenth and Nineteenth Centuries' *Eng. Hist. Rev.* LIX (1944) p. 238, and the discussion in J. D. Chambers, 'Enclosure and Labour Supply in the Industrial Revolution' *Econ. Hist. Rev.* 2nd ser. V (1952–3) pp. 325–7.

[2] W. G. Hoskins, *The Midland Peasant* (1957) pp. 164, 249–51.

[3] Olga Wilkinson, *The Agricultural Revolution in the East Riding of Yorkshire* (E. Yorks Local History Soc., 1956) p. 10.

and in the Vale of Pickering, as William Marshall the contemporary agricultural writer pointed out, it was the owners of houses with common rights attached to them, rather than the owners of farmland, who pressed for the enclosure of the commons.[1]

The consequences of enclosure for the village population we must now discuss, but in regard to the procedure of obtaining parliamentary authority and the re-allotment of the land we can say this: that in the opinion of most careful investigators it almost always worked fairly as between the various classes of proprietors; that it is not true that all small proprietors were opposed to it; and that in at least a proportion of cases the equitable claims of the squatters and the poor were taken into consideration. This was not perfect justice, but in an age of aristocratic government and exaggerated respect for the rights of property, it was not a bad approximation to it. Indeed, it may well be said that parliamentary enclosure represented a major advance in the recognition of the rights of the small man.

In the words of the Hammonds, 'enclosure was fatal to three classes: the small farmer, the cottager, and the squatter.'

Let us begin with the small farmer and see how he might be affected by the abolition of open fields and commons. It will simplify the discussion if we consider first the small farmers who owned their farms: the freeholders or owner-occupiers. (Strictly there was no clear division between owner-occupiers and tenant-farmers, for many small owner-occupiers rented some land, and they often rented more land than they owned; but we will ignore this complication for the moment.)

The main way in which the Hammonds supposed small owners to be adversely affected was through the burden of expenses involved in enclosure. Apart from mentioning some enclosures where the cost was unusually high, however, they produced no evidence that these expenses were in fact fatal to small owners. And when the matter is looked at closely it does not seem very probable that the cost of enclosure, by itself, forced very many small proprietors to sell out and decline to the status of labourers. It was only the really small owners, with holdings too small to be described as farms, who were likely to have found the expenses too great for it to be worth keeping their land: they depended on some supplementary or alternative occupation in any case, and they may well have found the occasion of the enclosure a good time to sell. Many of these, it must be remembered, were *absentee* owners who had let out

[1] W. Marshall, *Rural Economy of Yorkshire* (2nd ed. 1796) I pp. 51-2, 82.

their land, and not occupiers. Indeed, some of the sales of absentee owners must have consolidated the holdings of small farmers, and strengthened their position. In addition, many large landowners at this time were selling some land in order to raise money for enclosures and other improvements, or to clear up their debts. During the reign of George III private estate Acts sought by landowners in order to allow sales or freer use of land tied up in family settlements were nearly half as numerous as the enclosure Acts. No doubt a good deal of this land went into the hands of the gentry and smaller owners.[1]

The true small farmer, with his 30 or 40 acres, may have been faced with a total bill of anything from £30 to £250, according to the circumstances. In general, however, unless the enclosure was unusually costly, he would not have to find more than about £50 to £100, with the payments spread over a period of time. If we suppose that his land was reasonably fertile and that much of it had been open before the enclosure, then its value would probably have at least doubled in the enclosed state, and would have a market value of £25 or £30 per acre, in total say £900 for the whole farm. It follows that for an expenditure of £50 to £100 on enclosure his land increased in value from about £450 to £900. Thus there would be little difficulty in mortgaging his land to meet the enclosure costs, or alternatively he might raise the sum by selling off a half-dozen acres and still show a capital gain on the transaction. Or again, he might prefer to sell all his land at its improved value and use the capital to stock a really substantial farm as a tenant.

Of course, there is no doubt that some small owners who raised mortgages in order to meet enclosure costs, or to stock a farm newly converted from arable to pasture, in the end found themselves unable to pay off their debts and were forced to sell. Between 1760 and 1813, however, this seems unlikely to have affected large numbers of small owners for prices were rising and farmers were prosperous though the costs of enclosure were higher than early estimates suggested, and they tended to rise, especially in the war period. A study of 70 Parliamentary enclosures in the Lindsey division of Lincolnshire found that 82 per cent of owners receiving allotments were owners of less than 50 acres. In nine villages there were some 70 or 80 sales of land before or after enclosure, and of course there may have been others that cannot be traced. These sales, however, did not mean a permanent decline in the number of small owners in total, for the land tax assessments show that they were more than made up by fresh purchasers during the period of high prices at the

---

[1] Thompson, *op. cit.* pp. 213–14.

end of the century.[1] Where small owners eventually diminished after enclosure, as at Wigston Magna, it seems probable that other factors were also important, the rise in the burden of poor rates for instance, and the general long-run tendency towards larger and more efficient farms—a tendency which certainly did not begin with enclosure and did not end with it.[2]

Many small owners, indeed, had been bought out before enclosure, as a necessary preliminary to it. There were obvious advantages in simplifying the re-allotment of lands and making the farms larger and more compact if there were fewer owners to consider, and it was generally the case that the earlier enclosures occurred where the small owners were few or their lands had been gradually bought up by the larger owners over a period of years. It does not follow from this that small owners were always opposed to enclosure and likely to try to prevent it. Detailed evidence shows that it was frequently the case that enclosure proceeded smoothly in villages where there were enough small owners to have prevented it had they combined for the purpose.[3] In discussing open-field cultivation we noticed that it was quite common for the farmers to co-operate in exchanges of lands, the stinting of commons, and the arrangement of small enclosures. There seems no reason why they would not have welcomed a complete enclosure and the abolition of inconvenient and inefficient open-field holdings. Sometimes, indeed, the initiative did come from the farmers. Thus we have a Gloucestershire farmer suggesting an enclosure to Lord Hardwicke's agent, 'the generality of the Nation being soe much Improv'd by Inclosures.'[4] At Somercotes in the Lincolnshire marshland the small common right owners were so eager for enclosure that they were prepared to indemnify the commissioner against his unwittingly committing legal irregularities.[5] A preliminary survey of the number of acres held by each freeholder at Bromley in Kent in 1799, and the side he would probably take if an enclosure Bill were promoted, showed that 19 freeholders owning 2,177 acres would be in favour of the project,

---

[1] T. H. Swales, 'Parliamentary Enclosures of Lindsey' *Lincs. Arch. Soc. Reports and Papers* (1936).

[2] In nineteenth-century Wiltshire, for instance, the small farms declined as much in old-enclosed as in newly-enclosed parishes, and the greatest decline in owner-occupiers occurred in parishes enclosed by agreement in or before 1781. See R. Molland, 'Agriculture c.1793–c.1870' *V.C.H. Wiltshire* IV (ed. E. Crittall, 1959).

[3] H. G. Hunt, 'The Chronology of Parliamentary Enclosure in Leicestershire' *Econ. Hist. Rev.* 2nd ser. X (1957–8) p. 269; H. L. Gray, 'Yeoman Farming in Oxfordshire' *Quar. Jour. Economics* XXIII (1909–10) pp. 321–2.

[4] Glos. R.O.: D 214 E 18 (1735).

[5] Joan Thirsk, *English Peasant Farming* (1957) pp. 240–1.

and 15 others owning 1,246 acres would be against it. But the 'ayes' and 'noes' fell into no definite pattern: some of the large freeholders were thought to be against the enclosure, and some of the small ones for it.[1]

The individuals who stood in the way of enclosure were often the small *absentee* owners who had their land or common rights let out to the open-field farmers, and sometimes it was the larger owners of pasture land who opposed it because they benefited from a shortage of grazing in neighbouring parishes. At Flintham, for example, the shortage of pasture in the village prevented the farmers from keeping sufficient stock to manure their land, but as it was reported,

> It appears to be pretty difficult to gett any scheme of enclosure agreed on, it has been twice attempted by several of the Inhabitants, and was so this last Year, but was putt a stop to by Mr Disney who owns about £200 a year there, and if he had not putt a stop to it, its probable Mr Francis Molyneux might, who is owner of Kneeton the next parish, which is all enclos'd, and he lett his pastures to Flintham Inhabitants and has a Right of Common in Flintham . . .'.[2]

The evidence of the Land Tax Assessments shows that from the later eighteenth century up to probably about 1815 small owners were actually increasing in number and in acreage, even in some heavily-enclosed counties. The complete accuracy of this source, however, must be regarded as doubtful, and the figures for assessments tell us little about the size of farms or the total acreage, both owned and rented, in the hands of small men.[3] The broad conclusions of the Land Tax Assessments, nevertheless, are valid and important. They show that enclosure was not a very important factor in the survival of owner-occupiers, at least when prices were good, and that there was an extremely numerous class of very small owners whose land could not have been sufficient for a full-time holding. The figures also reveal, incidentally, how very numerous were the small absentee owners of farmland; and that the fewest small owners, both occupying and absentee, were to be found, as might be expected, in the villages enclosed earlier in the eighteenth century by agreement.[4]

[1] Kent R.O.: U310 E1.

[2] Notts. R.O.: DDTN 5/6 (1737).

[3] See G. E. Mingay, 'The Land Tax Assessments and the Small Landowner' *Econ. Hist. Rev.* 2nd ser. XVII (1964-5).

[4] See E. Davies, 'The Small Landowner, 1780-1832, in the Light of the Land Tax Assessments' *Econ. Hist. Rev.* I (1927) p. 111; J. D. Chambers, 'Enclosure and the Small Landowner' *Econ. Hist. Rev.* X (1940) p. 127; H. G. Hunt, 'Landownership and Enclosure 1750-1830' *Econ. Hist. Rev.* 2nd ser. XI (1958-9).

The increase in the numbers of small owners may have arisen in part from the recognition as freeholders, and taxpayers, of the owners of common rights, of copyholders, and even of squatters with encroachments on the waste. Small absentee owners found the high prices of the war period an inducement to sell, and we know too that large owners sometimes sold off part of their land to meet enclosure expenses. Tenant-farmers and existing owner-occupiers provided a ready market for small parcels of land at this prosperous period. The shrinkage in the number of small owners with the fall in prices at the end of the wars, and the fall in the area occupied by them, suggest that it was not enclosure but the level of prices in years of depression, and possibly the growth of alternative occupations outside farming, which were the important determinants of survival. According to Davies's figures for Cheshire, Derbyshire, Leicester-shire, Lindsey, Northamptonshire, Nottinghamshire and Warwickshire, the decline in the early nineteenth century was not very marked: in 1832 there were about as many, or more, small owners as in 1780, although the totals had fallen from the wartime peak. It is possible that in areas of light soils, where the competitive advantage of large farms was greater, the decline may have been considerably greater than Davies's figures suggest. At all events, at the end of the nineteenth century only about 12 per cent of the cultivated land was worked by its owners (excluding the land of larger proprietors currently in hand); and it is known that this figure was near 20 per cent about 1800, so that there was evidently a consider-able decline in the land occupied by small owners in the nineteenth century after 1815, and probably even after the end of active enclosure of open fields and commons.

Turning to the more general question of the decline of small farmers as a whole, there was undoubtedly a long-term tendency in favour of the consolidation of farms into larger and more efficient units. This tendency was encouraged by enclosure, but in no sense depended on it. The move towards larger farms was marked even in open-field villages, and the most drastic changes in the size of farms sometimes occurred before enclosure. At Wigston Magna the size of farms grew steadily throughout the seventeenth century, and the enclosure of the village in the eighteenth century 'did not directly create any greater inequality than had existed beforehand'.[1]

Farms grew in size for a whole variety of reasons: because except in certain specialized branches of farming large units were technically more efficient and more progressive; because the advances in husbandry

[1] Hoskins, *op. cit.* pp. 231–2, 253.

involving the cultivation of roots and legumes generally demanded fairly large acreages; because large farmers had the resources to withstand occasional bad years and periods of low prices; and perhaps not least because landowners took great care in selecting the tenants for their large farms, and saw that they had the capital and knowledge to farm successfully. The long-term tendency was therefore in favour of larger farms, and enclosure usually served only to accelerate the effect of factors unfavourable to small men.[1] The change was essentially a gradual one: there was no sudden or cataclysmic decline of small farms, and to speak of their 'disappearance' in the eighteenth century is absurd. By 1851 the advance of large-scale farming had indeed gone so far that farms of 300 and more acres occupied over a third of the cultivated acreage, while small farms and holdings of under 100 acres occupied less than 22 per cent. But the number of small occupiers was still very large—over 134,000 —as compared with 64,200 farmers of 100 to 299 acres, and only 16,671 farmers of 300 acres or more. The family farmer who employed no labour beyond that of his family was still very much in evidence. In 1831 of the 275,000 farming families in Britain, nearly half fell into the family-farmer category.[2]

Of course, enclosure itself might have a considerable impact on small farmers if it led to large-scale engrossing or encouraged a considerable change in the nature of the farming. Enclosure provided landlords with an excellent opportunity of consolidating small farms into large ones which were easier to manage and meant less outlay on building and repairs. But on the other hand, as landlords well knew, small farmers paid a higher rent per acre than did large ones, and owing to the considerable capital required to stock large farms suitable tenants for them were not easy to come by. Indeed, to a considerable extent the pace of enclosure was determined by the availability of farmers able and willing to change their methods and pay the higher rents landlords expected from enclosed farms. Consequently, many landlords preferred (or were obliged) to leave their small tenants undisturbed, and there was in fact a powerful convention among reputable landlords that they did not disturb tenants who were competent and paid their rents regularly.

The fact that many small owners rented some land in addition to their own complicated the situation. The availability of additional land clearly improved the efficiency and flexibility of freeholders and made their

---

[1] See G. E. Mingay, 'The Size of Farms in the Eighteenth Century' *Econ. Hist. Rev.* 2nd ser. XIV (1961–2).

[2] Clapham, *op. cit.* I pp. 450–1, II pp. 263–4.

post-enclosure survival more probable. But it was considered undesirable for landlords to have their land let in this way because of the tendency for freeholder-tenants to neglect the rented land in favour of their own. Consequently, some large owners no doubt saw enclosure as a good opportunity for reform, and consolidated their rented plots into substantial farms to be let to tenant-farmers only. Where this happened the acreage available to freeholders would fall and their post-enclosure prospects would diminish to that extent. How important this was in the survival of small freeholders it is impossible to say, but for many of them it may well have given rise to more intractable difficulties than did the more obvious but less serious problem of enclosure costs.[1]

Quite often enclosure was undertaken in order to extend the practice of convertible husbandry, and where the farms were too small for this, or the occupiers could not afford to marl their land or pay the high rents demanded after enclosure, some of them might after a period of years give way to larger farmers. In the west and central Midlands especially— in Leicestershire, Warwickshire, Worcestershire, Northamptonshire, Huntingdonshire and Buckinghamshire—there was a strong tendency for enclosure to extend the area under grass. The low acreage of many of the post-enclosure farms, unsupported by common pasture, made it difficult or impossible for some small men to succeed in convertible husbandry, and sometimes in fattening or dairying. In the Vales of Evesham and Berkeley corn growing and population were reduced by the conversion of deep-soiled arable ground to permanent pasture; in Lincolnshire the technical advantages of amalgamation of marshland with upland farms made the large farms larger, partly at the expense of smallholders;[2] in chalkland areas, as in Wiltshire, Hampshire and Dorset, enclosure reinforced an old-established tendency towards expansion of arable cultivation at the expense of downland sheep pastures. In this development, too, the small men found conditions against them: reduced pasture made it difficult to graze their flocks, and they were obliged either to combine flocks for folding purposes, or to resort to artificial grasses and roots and all the capital and labour required in alternate husbandry in order to feed their flocks.[3] Nevertheless, the post-enclosure changes did

[1] See the discussion of this aspect of the problem in V. M. Lavrovsky, 'Parliamentary Enclosures in the County of Suffolk (1797–1814)' *Econ. Hist. Rev.* VII (1937) pp. 193, 207–8, and 'Tithe Commutation as a Factor in the Gradual Decrease of Landownership by the English Peasantry' *Econ. Hist. Rev.* IV (1933) p. 273.

[2] Joan Thirsk, *op. cit.* p. 242.

[3] See E. L. Jones, 'Eighteenth-Century Changes in Hampshire Chalkland Farming' *Ag. Hist. Rev.* VIII 1 (1960) p. 10; T. Davis, *General View of Wiltshire* (1794) pp. 49, 177.

not always put small men at a disadvantage. Where the soils were sufficiently varied to allow mixed farming and specialization to succeed the small man could prosper, and there is good evidence of stability in the farming population even in areas where arable was converted to permanent pasture, as in the north-eastern districts of Leicestershire and in Rutland.[1]

It is important to remember, however, that the impact of enclosure on husbandry practices was very much less than has often been supposed. Changes in agriculture rarely come about suddenly. The movement towards large arable farms in East Anglia, towards permanent pasture and ley farming in the Midlands, and towards the extension of arable at the expense of sheep in chalk areas of the south were all in train in the seventeenth century and earlier, and certainly long before the period of heavy enclosure after 1760. Enclosure generally accelerated or intensified trends towards more productive farming, but it was not always the initiating force of these trends. Therefore it is not to be expected that changes in the size of farms or in their occupiers should necessarily proceed very much more rapidly after enclosure than before it. And in some areas, as in Cheshire, Wales and the northern counties where the enclosure was mainly of waste, there was an increase rather than a decline in the number of holdings.

Of course, the very changes brought about by enclosure, the opportunities which it might provide for adopting different and more productive methods of cultivation, could have a very stimulating effect on the farmers, changing their whole outlook. Dr Thirsk found that in Lincolnshire enclosure 'roused ambitions in the ordinary farmer for the first time, and that the fresh opportunities, suddenly opened up, brought into action stores of human energy never previously tapped. The psychological effect of change doubled and trebled the force of the original stimulus, with the result that people were willing to go beyond the economic limit in expending effort and money on their farms. The steward of the Duchy of Cornwall lands in the soke of Kirton-in-Lindsey twice commented on the large, almost foolhardy, outlay of tenants, who had no certainty of tenure of their holdings. An occupier of land at Heapham had carried manure seven miles to cover his new enclosure, while the lessee of Kirton-in-Lindsey manor was enclosing and building in 1796 "with a spirit almost unequalled considering his term was a short one for an undertaking of such great magnitude".'[2]

However, there is also a good deal of evidence that in many cases enclosure had little or no effect upon the character of the farming. This

[1] See Chambers, *op. cit.* pp. 328–31, 335.
[2] Joan Thirsk, *op. cit.* pp. 296–7.

might be because even in the open fields the land might be cultivated 'in severalty', i.e. without common rights, or because the surviving area of open fields was very small. An Enclosure Act might be invoked merely in order to incorporate waste land into the farms or to consolidate the existing holdings and closes for easier working. Enclosure did not necessarily lead to changed or better farming—and especially was this so if the soil did not lend itself to the cultivation of turnips or clover, or if the farmers were conservative. Marshall was one of a number of authorities who held summer-fallowing to be essential on heavy soils, and it is clear from the county *Reports* that bare fallows continued in many different areas. In Cambridgeshire Vancouver found enclosed parishes still following the rotations of open-field parishes, and Young noticed the same thing in Lord Chesterfield's new enclosures in Buckinghamshire. In Lincolnshire and Bedfordshire Stone and Batchelor saw only few signs of improvement, and at Knapwell in Northamptonshire Young found the husbandry so bad that he was at a loss to know why the proprietors had troubled to enclose. And in Wiltshire Davis believed that enclosure had led to deterioration, because some farmers had extended their arable without adopting new rotations while at the same time allowing their sheep flocks to decline.[1]

Finally, it may be argued that what really counted in the survival of small farmers were the levels of prices and costs. As we have remarked before, the rising prices between 1760 and 1813 must have made it possible for many small men to carry on who would have failed in the conditions prevailing before 1750, and who perhaps actually did fail in the depression after the Napoleonic Wars. It seems likely that their costs per unit of output were higher than those of large farmers, partly because they paid a higher rent per acre and bore a relatively heavier burden of poor rates, and partly because they lacked the economies of scale, the technical efficiency, capital resources and flexibility which helped the larger farmers. Thus, when prices fell their higher costs might prove fatal, as was the experience, for example, of the unfortunate farmers of the unimproved heavy clays.

We come last to the lowest level in rural society, the cottagers and squatters. 'The effect on the cottager can best be described by saying that

[1] W. Marshall, *Review of Reports to the Board of Agriculture* (York, 1808–18), esp. III pp. 49–53, 618; IV pp. 610–18; V pp. 381–2; A. Young, *Tours in England and Wales* (1932) p. 204; *Eastern Tour* I p. 24; T. Stone, *General View of Bedfordshire* (1794) p. 26.

before enclosure the cottager was a labourer with land, after enclosure he was a labourer without land.' Thus the Hammonds. There is indeed a great deal of truth in the Hammond's brief summary, for access to commons and waste may have played an important part in the economy of many cottagers. Such access might make it possible for them to keep pigs, a cow or some geese, to gather fruit and firewood, and in the case of the squatters to find a place for their dwelling, such as it was. The removal of this prop of the labourers' existence was undoubtedly a factor in the increasing poverty which characterized much of the countryside in the later eighteenth century and after.

However, we must be careful not to exaggerate the extent or the importance of the loss. There is some evidence to support those contemporaries who held that the commons were of little real advantage to labouring men, for when in some enclosures land was set aside for labourers' cow pastures or vegetable gardens it was sometimes difficult to find cottagers to take them. On the other hand, there is contrary evidence that such allotments were greatly appreciated by the labourers, and were instrumental in keeping them off the poor rates. Perhaps much depended on the local circumstances, the amount of employment available, the situation of the allotments and the nature of their soil, and so forth. In any case, it must be remembered that even before enclosure the majority of cottagers had no rights of common. Such rights did not belong to every villager but were attached to open-field holdings or certain cottages, and only their owners or occupiers were certainly entitled to make use of them.

The legal owners of common rights were always compensated by the commissioners with an allotment of land. (The *occupiers* of common right cottages, it should be noticed, who enjoyed common rights by virtue of their *tenancy* of the cottage, received no compensation because they were not, of course, the owners of the rights. This was a perfectly proper distinction between owner and tenant, and involved no fraud or disregard for cottagers on the part of the commissioners.) Unfortunately, the allotment of land given in exchange for common rights was often too small to be of much practical use, being generally far smaller than the three acres or so required to keep a cow. It might also be inconveniently distant from the cottage, and the cost of fencing (which was relatively heavier for small areas) might be too high to be worth while. Probably many cottagers sold such plots to the neighbouring farmers rather than go to the expense of fencing them, and thus peasant ownership at the lowest level declined.

Most labourers who made use of a common had access not by right but by custom. Customary rights to enjoyment of the common were sometimes recognized by the Commissioners, but quite often not. In any case, the common was usually a very limited benefit. In Clapham's opinion the right to keep a cow was probably rare, and Young's famous outburst contained much exaggeration: 'by nineteen Enclosure Acts out of twenty, the poor are injured, in some grossly injured... The poor in these parishes may say *Parliament may be tender of property; all I know is I had a cow, and an Act of Parliament has taken it from me.*' 'These parishes', commented Clapham, 'were not the nineteen out of twenty.' In fact there were some places where the poor were left with cow pastures. At Sutton Cheney in Leicestershire, for instance, provision in the shape of 'a sufficient quantity of land' (44 acres for 13 cow-keepers), was made 'for all those who had cows before'. In general, however, there is strong evidence that allotments and cottage gardens were scarcest in areas of recent enclosure, and particularly in corn country. Nevertheless, over England as a whole the majority of labourers, 'probably the great majority', had either a garden or a patch of potato ground.[1]

To some extent the loss of commons might be compensated, however, by an increase in the volume and regularity of employment after enclosure. Where the area of farmland was considerably expanded by the cultivation of commons and waste there would obviously be an increased demand for labour, and the work of the enclosure itself, in the making of fences and hedges, and in the laying down of new roads and the building of new farmhouses and barns, also created much employment. In the newly-cultivated fenlands of Lincolnshire, for instance, there was for many years an acute labour problem, and labourers had to be brought from old villages five or six miles away.[2] Furthermore, the extension of alternate husbandry and leys required more labour for the work of drilling and hoeing roots and the management of the beasts fed on the legumes and artificial grasses. Heavier crops required more hands in harvesting and in the winter occupations of threshing and winnowing. The processing of agricultural produce, the milling of flour, the malting of barley, the drying of hops, the making of cheese, the curing of bacon and the tanning of leather, and all the rural trades and crafts serving the farmers were expanded as the cultivated area and agricultural output rose.

It should be emphasized that in the eighteenth century, with the exception of the late innovation (and only gradual adoption) of the threshing machine, the improved methods of farming were *not* labour-saving

[1] Clapham, *op. cit.* I pp. 113–21.
[2] Thirsk, *op. cit.* p. 217.

(although it is probable that the labour required per unit of output was reduced). And in so far as enclosure encouraged the rise of better farming and an expanded acreage it must have greatly increased the supply of rural employment. Only where permanent pasture increased at the expense of arable was the labour requirement likely to fall off. In consequence there was in fact no general exodus of unemployed rural labour, pauperized by enclosure, to seek work in the industrial centres. The population of the numerous Nottinghamshire villages affected by enclosure rose only slightly less fast in the early nineteenth century than did the villages dominated by mining and textile manufactures. The census figures also confirm that agricultural employment was expanding, not declining: in 1831, for instance, 761,348 families were reported as employed in farming as compared with 697,353 families 20 years earlier.[1]

In view of this evidence it is difficult to understand the attacks that contemporaries made on enclosure as the cause of unemployment and depopulation in the countryside. Before the censuses of the nineteenth century, however, there were no reliable statistics to be obtained; and enclosure, it seemed, was a subject on which few writers could avoid making statements completely unsupported by evidence, or on which they could eschew rancour and hyperbole. Dr Price, one of the most bitter opponents of enclosure, claimed that its effects were visible in the half-empty and decayed churches, the ruined houses and unused roads of East Anglia. His remarks drew the fire of no less an enthusiast than Arthur Young. Rural improvement, answered Young, in the work of enclosing, marling, dunging, ploughing, turnip-hoeing, etc., had given more work than ever before, and particularly in Norfolk, the most improved county. The annual influx of Scots and Irish seeking harvest work, and the numerous wastes and heaths brought into production proved his point: 'INCLOSURES alone have made these counties smile with culture which before were dreary as night.' As for the half-empty churches, they had been built too large in the first place (which was true), while ruined houses were largely the result of the poor laws which induced landlords and farmers to pull down cottages and hire men from other parishes. This last contention also had some truth in it, for the Law of Settlement did lead to a growth of 'closed parishes' where to keep down the poor rates newcomers were refused a settlement.[2]

Even the poets were locked in combat on the subject. In celebrated

[1] J. D. Chambers, *op. cit.* pp. 322–4.
[2] A. Young, *Political Arithmetic* (1774) pp. 96–104, 145–8, 150, 155, 289–95, 322–31.

verse Goldsmith mourned his *Deserted Village*, and in the next century John Clare declaimed:

> '*Inclosure, thou art a curse upon the land,*
> *And tasteless was the wretch who thy existence plann'd.*'

But the Reverend John Tyley, a Northamptonshire vicar, described in Latin verse how the poor people, who first opposed enclosure and pulled down the fences, came to value its lasting benefits. To him the huge 'unbroken tracts' of the pre-enclosure landscape 'strained and tortured the sight', and the new hedgerows created shelter, shade and bird-song, and the 'beautiful sight' of bountiful harvests.[1] John Dyer, the Carmarthenshire painter-poet, also dwelt upon the disadvantages of the open fields, while Tennyson made his northern farmer claim that the wild, bracken-covered waste, support for not even one poor cow, was converted into rich pasture bearing 80 beasts. Thus wrote Dyer:

> '*Inclose, inclose, ye swains!*
> *Why will you joy in common field, where pitch,*
> *Noxious to wool must stain your motley flock*
> *To mark your property?... Besides, in fields*
> *Promiscuous held all culture languishes;*
> *The glebe, exhausted, thin supplies receives;*
> *Dull waters rest upon the rushy flats*
> *And barren furrows; none the rising grove*
> *There plants for late posterity, nor hedge*
> *To shield the flock, nor copse for cheering fire;*
> *And in the distant village every hearth*
> *Devours the grassy sward, the verdant food*
> *Of injur'd herds and flocks, or what the plough*
> *Should turn and moulder for the bearded grain...*
> *And too, the idle pilf'rer easier there*
> *Eludes detection, when a lamb or ewe*
> *From intermingled flocks he steals; or when,*
> *With loosen'd tether of his horse or cow,*
> *The milky stalk of the tall green-ear'd corn,*
> *The year's slow rip'ning fruit, the anxious hope*
> *Of his laborious neighbour, he destroys.*'[2]

[1] J. Tyley, 'Inclosure of Open Fields in Northamptonshire' (translated from the Latin by D. Holton), *Northants Past and Present* I (1951).

[2] A. H. Dodd, *The Industrial Revolution in North Wales* (Cardiff, 1933) p. 53; Lord Tennyson, 'Northern Farmer (Old Style)', verse X (*Works* (1892) p. 230).

But whatever the merits of the controversy, both sides recognized that poverty was increasing in the countryside. Even the protagonists of enclosure were obliged to admit this unpalatable fact. Arthur Young, although he happily found a number of instances where enclosure had not been accompanied by rising poor rates, was dismayed to see that in general the poor rates in enclosed parishes, despite increased employment, had kept pace with those in other parishes. After a tour of villages in the south of England he arrived at the conclusion that the phenomenon arose from the failure of the commissioners in most enclosures to allot part of the commons or waste to the poor. Properly supervised allotments, he argued, had been shown greatly to benefit the poor and were effective in reducing the poor rates. Some landlords, the Duke of Bedford, the Earl of Egremont and Lord Hardwicke, for example, had done much good by providing cottage gardens and prizes for the best cultivation. But he found that in 25 of the 37 enclosures he examined the lot of the poor had deteriorated.

Young noticed that in many recently-enclosed villages there were still areas of waste left uncultivated, and he believed that the cheapest way of providing for the poor would be to put labouring families on this land with a cottage and three acres apiece. They should be allowed to keep this allotment indefinitely so long as they never asked for relief from the parish. To do this, he estimated, and provide a cow, pig, seed and implements, would cost about £50 per family. This might seem a large capital outlay, but in the long run it would prove economical for it cost £60 to keep a family of five in a workhouse for a year, and £20 a year for the same family maintained on outdoor relief. Property, he always held, gave the poor the incentive to work hard, be frugal and save, and was effective in keeping them off the parish. And for illustration he quoted 48 parishes in Lincolnshire and Rutland, where there were as many as 753 cottagers with cow pastures and nearly 1,200 cows between them. It had been found that these cottagers did not come to the parish for relief, and they were acknowledged by the farmers to be the most hard-working of labourers. Indeed, the poor rates in these parishes were less than half the average elsewhere.

There was the problem, however, that in many villages, and especially in those enclosed before the era of parliamentary enclosure, no waste land was now left. Here the poor could not even help themselves: they could not find land to build their own cottage (which could be done for a little over £20, paid back to the carpenter over 15 years); nor could they cultivate a vegetable patch, nor keep a pig. It was to this situation that his most famous quotation relates: 'Go to an alehouse kitchen of an old-enclosed country, and there you will see the origin of poverty and the poor-rates. For whom are they to be sober? For whom are they to save? For the parish? If I am diligent, shall I have leave to build a cottage? If I am sober, shall I have land for a cow? If I am frugal, shall I have half an acre of potatoes? You offer no motives, you have nothing but a parish officer and a workhouse. Bring me another pot.' It is important to notice that this was said in reference to *old-enclosed* villages, and not to enclosure in general.

Where no waste land was left, and kind landowners willing to let land to the poor were lacking, Young suggested that the parish itself should buy some land, borrowing the money on the security of the rates, and should fence and stock it and let it out to the poor. Young's scheme was only a partial solution, of course. It might have prevented much misery and degradation, but it could not overcome the effects of rising numbers. He himself noted that families settled on allotments revealed an alarming tendency to propagate, 'so that pigs and children fill every quarter', thus intensifying the very problem he sought to solve.[1]

In practice it was not at all easy to distinguish the influence of enclosure among the various factors affecting poor-law expenditure. With the growth in the countryside of such industries as textiles, mining and iron manufacture, the level of employment was affected by booms and slumps in trade. In some areas, as in East Anglia, Sussex and the Forest of Dean, the decline of the obsolete textile, coal, and iron industries gave rise to growing unemployment. Even in villages still entirely agricultural harvest failures or the growth of permanent pasture could have serious effects. Marshall held that 'agriculture occasions very few poor, on the contrary it provides them almost constant labour. It is only the blind, the extreme old, the very young children and idiots which become chargeable in a parish purely agricultural.' He was surprised to find that in some places this was not true.[2]

We see now that the fundamental factor was the great upswing of population. Population increase, which became far more rapid in the later eighteenth century than hitherto, was expanding the labour force at a rate faster than agriculture could absorb it, and the growth of numbers, of landless and sometimes unemployable labourers, was

---

[1] A. Young, *An Inquiry into the Propriety of applying Wastes to better maintenance and support of the Poor* (Bury 1801) pp. 21–5, 37 ff.; *General View of Lincolnshire* (1799) p. 462.

[2] W. Marshall, *Review of Reports to the Board of Agriculture* IV p. 203; V pp. 118, 461.

observable both in enclosed and the still open villages. It was this natural phenomenon, the orgins of which are still obscure, which lay at the bottom of unemployment and the rising poor rates. Of course, there had always been a good deal of rural poverty: in Leicestershire in 1670 there was already a large proportion of villagers too poor to pay the hearth tax.[1] There was also at the end of the seventeenth century already a high proportion of landless labourers. The changes of the eighteenth and early nineteenth centuries added to this rural proletariat, but not to the extent that might be supposed: about 1690 there were nearly as many as two landless labourers to every occupier; and in 1831, after nearly a century and a half of enclosure and engrossing, there were still only five landless labourers to every two occupiers. More precisely, as Sir John Clapham pointed out long ago, the proportion of landless to occupiers rose only from 1.74:1 to 2.5:1.[2] Much of this increase came from the growing numbers of surviving younger sons with no land to inherit; population growth was the main factor in the increase in a landless, as well as a partially workless, labour supply in the countryside.

The increase in population was spread fairly evenly over the country, but in areas of developing industries there was migration away from the agricultural villages, able-bodied workers attracted by better pay and wider opportunities of employment moving over distances of 20 or 30 miles to the industrial towns and villages. In consequence, the effects of population growth on rural unemployment and poverty were much more marked where alternative industrial occupations were not available. The statistics of poor relief bear this out, showing that the distribution of poverty was not related to the extent of recent enclosure but to the availability of work outside farming. The per capita expenditure on poor relief was higher in some counties virtually unaffected by eighteenth-century enclosure than in others exposed to the full flood of the movement. According to Gonner's figures, Kent and Sussex, for example, spent as much or more per head on relief as did the most heavily-enclosed counties of Bedfordshire, Huntingdonshire, Leicestershire, Rutland and Northamptonshire; and much more was spent in heavily-enclosed counties that remained largely agricultural, like Berkshire, Wiltshire, Oxfordshire and Huntingdonshire, than in the heavily-enclosed but semi-industrialized Nottinghamshire and Leicestershire.[3]

[1] Joan Thirsk, *V.C.H. Leicestershire* II (1954) p. 228.
[2] J. H. Clapham, 'The Growth of an Agrarian Proletariat, 1688–1832' *Cambridge Historical Journal* I (1923) pp. 93–5.
[3] Gonner, *op. cit.* p. 448.

The same picture emerges if one looks not at poor law expenditure but at the proportion of population actually in receipt of relief. The counties little affected by enclosure, like Kent, Sussex and Hertfordshire, had as high a proportion of paupers as had the heavily-enclosed Northamptonshire, Leicestershire, Rutland and Nottinghamshire. And more detailed investigations within the heavily-enclosed counties have shown that whether the villages were enclosed recently by Act of Parliament, were old-enclosed by agreement, or still remained open, had little connection with the proportion of population receiving relief.

Returning to the importance of alternative employment in the distribution of poverty, it is very significant that it was not the heavily-enclosed Midlands but the old-enclosed, poverty-stricken counties of southern England, Kent, Sussex, Hampshire and Dorset, that were the scene of the last Labourers' Revolt of 1830. Undoubtedly low wages, unemployment and bad living conditions lay at the origin of the outbreaks of violence and destruction, but there was obviously no connection between the revolt and enclosure, as the Hammond's references to it would suggest.

To sum up, the effects of enclosure were rarely great or immediate. In some instances enclosure came as the last act of a long-drawn-out drama of rural change, and merely put *finis* to the story. In others it sometimes introduced, but more often accelerated, a similar story of change. As the result of enclosure improved farming spread more rapidly than would otherwise have been the case, larger and more efficient farms were more readily developed, and the long-run decline of the smallholder and cottager hastened and made more certain. Enclosure provides a leading example of the large gains in economic efficiency and output that could be achieved by reorganization of existing resources rather than by invention or new techniques. Enclosure meant more food for the growing population, more land under cultivation and, on balance, more employment in the countryside; and enclosed farms provided the framework for the new advances of the nineteenth century. But in our period enclosure did not affect the whole country, and even the limited area that felt its influence was not suddenly transformed.

Enclosure remains an important and indeed vital phase in English agricultural development, but we should be careful not to ascribe to it developments that were the consequence of a much broader and more complex process of historical change.

*Suggestions for further reading:*

| | |
|---|---|
| J. D. Chambers | 'Enclosure and Labour Supply in the Industrial Revolution' *Economic History Review* 2nd ser. V (1952–3). |
| J. D. Chambers | *Nottinghamshire in the Eighteenth Century* (2nd ed. 1966), Ch. 6–7. |
| E. C. K. Gonner | *Common Land and Inclosure* (2nd ed. 1966). |
| David Grigg | *The Agricultural Revolution in South Lincolnshire* (1966). |
| J. L. and B. Hammond | *The Village Labourer* (1912). |
| E. J. Hobsbawm and G. Rudé | *Captain Swing* (1969). |
| W. G. Hoskins | *The Midland Peasant* (1957). |
| J. M. Martin | 'The Cost of Parliamentary Enclosure in Warwickshire' *University of Birmingham Historical Journal* IX (1964). |
| J. M. Martin | 'The Parliamentary Enclosure Movement and Rural Society in Warwickshire' *Agricultural History Review* XV (1967). |
| G. E. Mingay | *Enclosure and the Small Farmer in the Age of the Industrial Revolution* (Studies in Economic History, 1968). |
| G. E. Mingay | 'The Size of Farms in the Eighteenth Century' *Economic History Review*, 2nd ser. XIV (1961–2). |
| A. J. Peacock | *Bread or Blood* (1965). |
| W. E. Tate | 'Opposition to Parliamentary Enclosure in the Eighteenth Century' *Agricultural History* XIX (1945). |
| Joan Thirsk | *English Peasant Farming* (1957). |
| M. E. Turner | 'The Cost of Parliamentary Enclosure in Buckinghamshire' *Agricultural History Review* XXI, 1 (1973). |

See also the useful 'Bibliography of Recent Work on Enclosure, the Open Fields, and related Topics', by W. H. Chaloner, *Agricultural History Review* II (1954) pp. 48–52.

# 5 Prosperity and Depression, 1750–1846

As an industry, agriculture is peculiarly subject to the effects of market conditions on the prices of its products. Its circumstances approximate very closely to those which economists have defined as necessary to the state of perfect competition, namely that there exists a large number of producers, each producing but an insignificant proportion of the total output of commodities which are homogeneous in nature; with the further circumstance, moreover, that until quite recently the producers were too numerous and too independent to be effectively organized into controlling their output. Before the 1930s farm prices depended on the relationship between the total supplies from thousands of farmers (including farmers overseas) coming on to the market and the food and raw material demands of the population. The inelasticity of the demand for the most important products, and the variations in the output of the farmers (influenced by the effects of weather on harvests, animal and plant diseases, natural calamities such as floods, and by the varying production decisions of the farmers themselves), naturally give rise to considerable price fluctuations, and for this reason farming has always been one of the most uncertain and unstable of all industries.[1]

Of course, some farmers were more dependent on market prices than others. Very small occupiers who produced partly for subsistence, or who sold a specialized or perishable product like milk, hops or vegetables on a

[1] The introduction in the 1930s of a protected home market, guaranteed prices, and marketing boards for certain products (e.g. wheat, barley and oats, potatoes, milk, bacon and pigmeat) was designed to create greater stability and prosperity in the industry.

sheltered local market, might be in a favoured position; but many small farmers were obliged by circumstances to rely largely on cash crops of wheat or barley, the prices of which were always fluctuating. In the eighteenth and nineteenth centuries, as farms grew in size and communications improved, the movements of prices influenced more closely the lives of a growing proportion of farmers, until probably none but the very smallest could afford to ignore them.

But what prices were of the greatest significance to farmers? Much of English farming was 'mixed', which is to say that the cultivation of arable crops was associated with the fattening of cattle and sheep, so that the great majority of arable farmers were concerned with the prices of a number of quite different products. While there were many pasture farmers who had little or no arable acreage, there were relatively few arable farmers who did not keep some sheep and cattle. Among the small farmers, too, a different kind of mixed farming often prevailed where soil and markets were convenient, their 30 or 40 acres yielding a little corn, some hay and potatoes, the produce of a few cows, and very likely pigs, poultry, eggs, hops and fruit. But despite the wide range of possible combinations there was also a considerable degree of regional specialization, which was associated with climatic and soil characteristics. Apart from the areas of highly diversified farming in the home counties and near large centres of population, the farmers of the eastern and southern counties depended mainly on arable products, particularly on wheat, and on barley, oats, hay and straw, although fatstock and wool were also important, and of course some parts of the east and south, as in Lincolnshire, Suffolk, Essex, the vale of Aylesbury and Romney Marsh, were given over to grazing and dairying. The western half of the English lowland area, south-western England and the border areas, were largely given over to grass, although enclaves of arable persisted, as in southwest Lancashire, parts of Staffordshire and Monmouthshire. But fattening and dairying were there the predominant branches of agriculture, and the farmers of the rich midland pastures depended ultimately on the demand for beef and mutton, and to a less obvious extent on the usefulness of wool, hides, tallow and other animal by-products. The dairymen had their milk, butter and cheese—the last commodity often travelling some hundreds of miles to market—as well as pork, hams and bacon, for the keeping of pigs was usually combined with dairying. Finally, the farmers of the upland areas, the Pennines, Yorkshire moors, Welsh hills and the remote north, depended largely on sheep and the rearing of the young stock for the fattening farms of the Midlands and south, although they also had useful stand-bys in wool, milk, butter and cheese sold in their local markets.

The principal market prices for the larger farmers were therefore wheat, barley and oats (the last being important more for horses than for humans); beef and mutton; milk, butter, cheese and pigmeat; and wool, tallow and hides. All these were subject to fluctuations in price as a variety of influences increased or diminished supply and affected the level of demand. Shortages of fodder, for example, or outbreaks of liver rot, might force a reduction of herds and flocks and so temporarily depress market prices; subsequently the reduced numbers of beasts coming to market would raise prices again. In general, however, it was the products of arable farming that were subject to the widest fluctuations, and the prices of meat, dairy produce and animal by-products were relatively stable. The large annual movements in arable prices were mainly due to the variations in size of the harvests and the consequent fluctuations in the amounts coming to market, while the demand for wheat, barley, rye and oats, being largely for bread, fodder and distilling, was for lack of alternatives, not subject to much variation. The demand for grain was therefore inelastic or unresponsive to price changes, while the supply varied considerably from year to year according to weather conditions. In consequence, when the harvest was good a much greater supply than normal met a stable demand and prices fell sharply, while in years of bad harvests shortages drove up prices because the demand was still almost as large. Of the grains wheat showed the greatest fluctuations, and oats were usually more stable in price than either wheat or barley. Weather conditions made a great difference to the wheat yield, not only in the harvest period but also in the sowing and growing season. In severely affected areas the produce might fall as low as a half of that of a good year. It seems that farmers generally profited from a rather indifferent harvest (but not, of course, a disastrously bad one), when prices more than compensated for the moderate yield, while their bitterest complaints were reserved for years of glut when prices slumped and their receipts hardly covered the high expenses of harvesting and threshing the bountiful crops.

The situation was complicated by the Corn Laws. These laws represented an old-established expedient for controlling somewhat the wide fluctuations of the grain market. In the eighteenth century the Corn Laws encouraged export of grain by a bounty when home prices were low, and allowed imports at low rates of duty when home prices were driven up by bad harvests. In so far as they encouraged export and restricted import, the Corn Laws must have kept home prices rather higher than they would otherwise have been, and probably they helped to stabilize prices and operated with reasonable fairness as between the interests of producers and those of consumers. This at any rate is the opinion of the expert students of the subject.[1] But the new Corn Laws of the period after 1790, and particularly of 1815, opened a phase of much higher protection for producers. The implications of this change in the principle of the Laws, and the eventual overthrow of the whole protective system, however, we take up later. It is worth noticing here that the wars of the eighteenth century do not appear to have had any remarkable effects on home prices; and even the more intensive warfare and economic blockade of the Napoleonic era had only a limited impact, felt in certain years when the home harvest was excessively deficient. As we shall show, it appears to have been mainly the coincidence of runs of very bad seasons with rapidly growing demand which accounted for the enormous prices of the war years after 1793.

In addition to the year-to-year fluctuations caused largely by weather factors, there were also discernible long-term trends which were of more lasting consequence. The rapid growth of population after about 1750 was the most significant of the factors affecting agriculture's long-term prospects. Before 1750 prices of both grain and animal products tended to rule low, mainly because of lengthy runs of good seasons, but also because the growth of the cultivated acreage and the gradual improvements in land-use, fodder supplies and rotations, raised output beyond the increase in consumption. After about 1750 there set in a long period of some sixty years in which prices generally tended upwards as output trailed after the rising consumption of a rapidly increasing non-agricultural population. The upward price trend was vastly distorted by runs of unusually poor seasons during the war years after 1793; and the post-war price fall was the necessary consequence of a return to more normal supply conditions with an acreage greatly expanded by the wartime enclosures (see fig. 2). When eventually the post-war readjustments were completed, the influence of the still-swelling population was still felt, and continued to push prices upwards (with the exception of wheat, already by the 1850s feeling the effects of free imports), until there came about the sea-change of the last quarter of the nineteenth century.

Industry grew alongside population, providing growing markets for farming's secondary products, wool, tallow and hides especially, and the improving living standards of a large proportion of the people meant that

[1] D. G. Barnes, *A History of the English Corn Laws* (1930) p. 29; C. R. Fay, *The Corn Laws and Social England* (Cambridge, 1932) p. 9.

the demand for meat and dairy produce rose more rapidly than that for bread corn, the mainstay of the poor. After 1850 meat production and dairying became markedly more profitable than grain, and as Caird pointed out, there was already a considerable margin between the rents which the grassland farmers of the western half of England could afford to pay and those of the arable farmers of eastern and southern England. This was not a very recent development for since the wars grass had probably been generally more profitable than the plough, but certainly the effects of rising living standards and of wheat imports after 1846 were to tip the balance even more heavily in favour of grass.

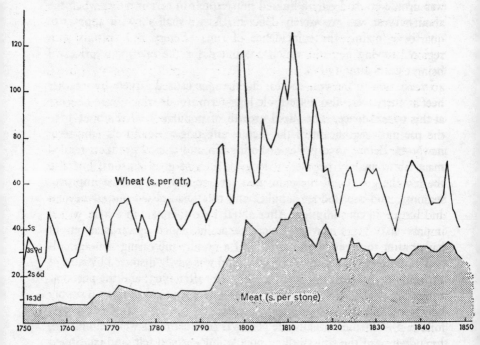

2  *Wheat and meat prices, 1750–1850*

Of course, the prosperity of farmers depended not only on prices but also on costs. Two important elements in costs, rents and wages, climbed with prices, but they generally showed substantial time-lags which left the farmers the chief beneficiaries of the rising price level after 1750. In the later eighteenth century, and particularly after 1793, rates and taxes rose heavily, the main factors being the rise in poor law and war expenditure. The price fall at the end of the wars found the farmers left high and dry with costs falling less quickly than prices. Hence the grumbling and complaints of depression, particularly from those farmers who found it difficult to farm more remuneratively, and who were hit by the monetary deflation and bank failures. By the later 1830s, however, conditions had improved: prices were better, and costs had been reduced, the farmers now feeling the benefits of the reform of the Poor Law, the commutation of tithes, and the abandonment or putting down to grass of high-cost marginal land.

Between 1750 and 1790, then, the general trend of agricultural prices was upwards. Discounting annual fluctuations, wheat rose from between about 26s. to 36s. per quarter about 1750 to the level of 45s. to 60s. per quarter in the late 1780s. The prices varied somewhat in the different regional markets, but the general trend was unmistakable, the price level being established at about 50 to 75 per cent higher in 1790 than it was 40 years earlier. Barley and oats rose in about the same proportion, and beef and mutton prices rose on balance about 25 per cent. The conditions at this time then appear to have favoured arable farmers rather more than the pasture farmers, and it seems likely that the greater proportion of newly-enclosed and newly-cultivated land came under the plough. The main factor in the rise of prices was the increase in demand arising from the growth of population, together with a decline in the frequency of good seasons and bountiful harvests. The rise in prices stimulated enclosure and brought about a large expansion of the cultivated acreage as well as improved farming methods, and presumably the poor harvests would have resulted in even higher prices had it not been for these developments.

The period of consistently large exports of grain came to an end after 1764, and in contrast the rise in prices attracted heavy imports of grain in 1767–8, the import of wheat in 1767 alone amounting to nearly 500,000 quarters. A period of precarious balance between import and export followed, with some occasional years of large imports. Such years were 1774 and 1775, and again in 1783, while exports were still considerable in 1776 and 1779–80. Indeed, the decade of the 1770s saw rather larger net imports than did the following decade, when the net balance of wheat imports was quite small, averaging only about 174,000 quarters a year. Although it was becoming clear that England had ceased to be a grain exporter, her usual import requirements amounted to only a tiny proportion of the total supply and were not yet critical. The rise in prices did lead, however, to a reconsideration of the Corn Laws; and by prohibiting export of wheat when it was at or above 44s., while allowing import at a nominal duty of 6d. when it was at or above 48s., the new Act of 1773 leaned more towards the consumer than the producer. The subsequent Act of 1791, however, was markedly more protective of the producer, in that it raised the limit for low-duty imports to 54s., and when wheat was under 50s. imports were to pay 24s. 3d. per quarter. However, these changes were of little real importance for wartime conditions meant that the Corn Laws were largely inoperative.

After 1750 there was a time-lag before rents rose and generally there were no sharp rent increases before the 1760s or 1770s. On land that was unaffected by enclosure, rents rose by about 40 to 50 per cent during the 40 years after 1750, most of the rise coming in the last 20 years of the period. Where land was newly enclosed, however, the rent increases would generally be considerably greater, of course, because of the consequent changes in land-use and greater productivity of the soil. There were no large increases in wages or parish rates until fairly late in the period, and there is clear evidence from many sources that farmers were much more opulent than hitherto, drinking their port wine 'by the pipe', sending their daughters to boarding schools, furnishing their drawing rooms with pianofortes and their dining rooms with 'elegant plate', and generally flaunting their new-found wealth to the disgust and scorn of their betters.

Farming prosperity was broken only by some occasional regional difficulties of spoilt harvests, fodder scarcities, and outbreaks of animal disease, a sharp and fairly general price fall in 1779–80, and an extremely severe winter in 1788–9 with turnip shortages and severe outbreaks of liver rot among sheep in the following autumn and the spring of 1790. Over the period as a whole, however, favourable prices encouraged the remarkable increase in both landlords' and tenants' investment and the widened enthusiasm for improved methods of farming which were discussed in earlier chapters.

With the wars of 1793 to 1815 the steady development and the gradual upward rise in prices of the period between 1750 and 1790 came to a sudden end. There was a sharp upswing of prices to quite unprecedented levels, accompanied by violent fluctuations which went far beyond the experience of that generation of producers and consumers; and the phenomenon was followed by a hectic scramble to enclose and expand cultivation.

In the early 1790s wheat hovered about the level of 48s. to 58s. per quarter. In 1795 it shot up to over 90s. (averaging in 1795–6 over 76s.), fell back to about the pre-war level, and then in the first years of the new century rose to even greater heights, the average price for 1800 mounting to 113s. 10d, and for 1801 to 119s. 6d. These fantastic prices could not be maintained but the downswing never fell to pre-war levels, and again in 1810, 1812 and 1813 the annual average was over 100s., that for 1812 being 126s. 6d. In only six of the 23 years between 1793 and 1815 did the average price of wheat fall below 65s., and in only ten of the years did it fall below 75s. (see fig. 2, page 110).

The story is similar with barley and oats, although barley prices were influenced by the periodical prohibitions on the use of grain in distilleries. Before the war barley was running at about 26s. per quarter but in 1795–6 it was up to nearly 38s., and in 1801 it reached the fantastic price of 90s. 7d.—twice the pre-war price of wheat. As with wheat, barley experienced great fluctuations, falling below its pre-war price in 1802–3 and rising to over 50s. in 1809–10 and to nearly 80s. in August 1812. Oats similarly ran up from the low levels of about 20s. per quarter to peaks of over 40s. and even 50s. in the bad years. For the grain producers the great heights of the war period stood out like a watershed between the rising foothills of the later eighteenth century and the subsiding plains of the long difficult years after 1820.

Meat prices, too, rose greatly with the poor seasons and dear fodder of the war years. Severe fodder shortages and outbreaks of sheep rot had already occurred in the five years preceding the war, and from the pre-war level of about 3s. to 4s. per stone, beef and mutton rose sharply and continued to climb until 1802, when prices were up by between 70 and 100 per cent. A period of more moderate prices followed, with a renewed climb 1808 to 1813 to the peaks of 6s. to 8s. 6d. per stone in December 1813, representing a total rise of over 100 per cent on pre-war levels. Then followed a sharp drop which continued until bottom was reached at 4s. or less in 1817 (see fig. 2, page 110). Between 1800 and 1815 the average price was about 5s. 3d. per stone, with mutton prices running rather higher than those of beef.

Weather conditions and the consequent effects on home supplies coming to market, rather than difficulties in obtaining imports, appear to have been the main factors in this strange wartime behaviour of prices. There were few good seasons, and it happened that the bad years occurred in long runs: this meant that in the absence of reserve supplies of grain, and with the reduced sizes of flocks and herds, prices built up to famine levels. The effects on prices of the peculiar weather circumstances were abetted by the inflationary consequences of government war finance and the over-issue of notes by country banks; and it was the combination of these factors, much more than high freight charges and Napoleon's continental blockade, that explains the great food crises of 1795–6, 1800–1, and 1809 to 1812. Indeed, in these years imports were greater than ever before.

The first big wartime upswing came in 1795. This infamously bad year—the year of Speenhamland—resulted from one of the three worst winters of the whole eighteenth century, and together with the unusual cold of the following summer produced a harvest a fifth or a quarter less than normal:[1] even in late June there were such sharp frosts that 'Thousands of new shorn sheep that were left out in the fields died of the Extreme cold'.[2] But unlike the two previous famine years of 1709 and 1740 the harvest failure of 1795 followed on a deficient harvest in the preceding year. Although in the war period the productive capacity of agriculture was constantly being expanded by the extension of the culti-vated acreage, and possibly by some conversion of pasture land to arable, this was more than offset by the general inclemency of the seasons, by the need to supply the armed forces, and of course, by the growing demand of a rapidly increasing population.

It happened that the whole of the period from 1795 to 1800 was remarkable for bad or indifferent seasons. In 1796 the harvest was poor. The summer of 1797 was held to be the wettest 'ever remembered this age', and the following winter was also very wet, with numerous out-breaks of sheep rot. The winter of 1798–9 was a severe one: 'the autumn was very unfavourable for wheat sowing; severe frosts set in early and heavy falls of snow with rapid thaws' destroyed a great deal of the turnip crop. The harvest of 1799 was spoilt by heavy rain and early frosts, and the autumn was again unfavourable for sowing wheat: 'we were so over-whelmed with Torrents of Rain day after day, and week after week, that it was completely out of the power of vast numbers of people on heavy Clay Soils to sow half their regular Quantity'; the following winter was similar to the preceding one, 'marked by early severity of frost, with alternations of rapid thaws; and the spring of 1800 was exceedingly wet.'[3]

The first few years of the new century saw much improved weather conditions and consequently increased home supplies and a welcome fall in prices, but from 1808 to 1812 there was another disastrous run of five

[1] See Mancur Olson Jr., *The Economics of the Wartime Shortage* (Durham, North Carolina, 1963) p. 50.

[2] Quoted by E. L. Jones, *Seasons and Prices* (1964) p. 153.

[3] *Ibid.* pp. 153–6; T. Tooke, *History of Prices* I (1838) pp. 213, 216.

bad seasons. Both 1808 and 1809 experienced wet summers with spoilt harvests and serious outbreaks of sheep rot; 1810 was almost wholly bad with winter cold destroying the grain in the ground and sheep rot prevalent in a cold and wet summer; 1811 again saw the harvest greatly reduced by the combination of a cold winter and a wet summer, and 1812 fitly crowned this dismal series with a deficient harvest which again sent grain prices up to famine heights.[1]

The wartime harvests necessarily meant an increased need for imports of grain, which were subject to enemy restrictions, shipping losses and high freight charges. The imports of wheat rose sharply and reached unprecedented levels in 1796, 1800–1, and 1810. The net imports of wheat totalled 854,000 quarters in 1796; 1,243,000 quarters and 1,397,000 quarters were imported in 1800 and 1801 respectively, and 1,491,000 quarters in 1810. Political difficulties arising from the war considerably obstructed the import of foreign grain, and the dearth of 1800–1 was intensified by the Russian embargo on British shipping and the Danes' closing of the Sound. Napoleon's continental blockade also had some effect on supplies between 1806 and 1809, and again in 1811–12 when the home harvests were perhaps not much above half the normal size. In 1810, however, Napoleon was faced with a glut of corn in France following a bumper harvest there in 1809. In order to raise French prices and keep his farmers quiet, but with the additional object of draining bullion from England to weaken her internal credit and her ability to subsidize her allies, Napoleon allowed large quantities of grain to be shipped from France to England under licence, thus moderating the English scarcity during a very critical food crisis.[2] The price of such imported grain of course rose considerably in transit due to high freight, insurance and licence charges, and Tooke in his *History of Prices* estimated that these totalled between 30s. and 50s. a quarter.[3]

Over the whole war period the net wheat imports averaged a little over 500,000 quarters a year compared with under 200,000 quarters in the 1780s. England thus became more definitely dependent upon foreign grain, three-quarters of which came from the Baltic area and was liable to interruption by unfriendly powers, while the remaining quarter came from the United States. Although much larger than before the war, the grain imports were, however, still small in relation to home production. It is calculated, for instance, that even the imports of 1810—the highest

[1] Jones, *op. cit.* pp. 158–9.
[2] Olson, *op. cit.* pp. 60–2.
[3] *Op. cit.* p. 327.

of the war period—represented only about nine weeks' consumption for the British population, and the complete loss of these supplies, although it would undoubtedly have caused intensified distress, would not have been an intolerable burden.[1] In any case, Napoleon could not prevent Britain from importing grain from Ireland and America, and his continental system, far from watertight, leaked copiously on the many miles of European coastline where the numerous smugglers could operate.[2]

Nevertheless, in Britain the recurring food crises and acute labour unrest led to renewed interest on the part of politicians, agricultural experts and pamphleteers in the causes of the very high prices and serious shortages. High food prices had been an occasion of rioting throughout the eighteenth century, and of course such disturbances became even more common in the years of wartime scarcity and distress. It was not unusual for the rioters to take over control of the market place, intercept supplies on the way to market, and sell them off at their own prices. Such direct interference in the normal course of trade was extremely widespread in the famine year of 1795. In Newcastle and Carlisle, for instance, the townspeople seized grain and other provisions and sold them at low prices, while the country folk often acted to prevent loads of grain from being sent off into the towns.[3] Landlords and large farmers were suspected of reaping undue gains out of the misery of the poor, and it was darkly said that in former times of scarcity the hanging up of a farmer or two had been sufficient to bring down prices. There was even greater hostility towards corn factors, millers, bakers and all middlemen who were believed to be guilty of rigging the market, and in 1800 the successful prosecution of a London corn dealer for the ancient offence of 'regrating' (buying and re-selling corn on the same day) was sufficient to spark off severe riots.

In each of the three great crises, 1795-6, 1800-1, and 1809 to 1812, the unrest forced the government to intervene with bounties on imports (a strange reversal of the export bounties of not so long before), and restrictions on consumption. These included the use of grain in distilling and in making starch and hair powder. Attempts to encourage a voluntary reduction in wheat consumption by the use of substitutes for bread such as potatoes and rice, adopted as an example to the poor by members of the wealthy classes, seem to have failed, as did also efforts to get the poor

[1] Olson, op. cit. p. 65.
[2] F. Crouzet, L'Économie Britannique et le Blocus Continental II (Paris, 1958) p. 571.
[3] See E. P. Thompson, The Making of the English Working Class (1964) pp. 65-6.

to give up their white bread for one of mixed or inferior flour.[1] White bread had become a symbol of a better standard of living for some three-quarters or so of the population, and these were not so poor as to readily go back to the indignity of brown bread. In Nottinghamshire the wealthy farmers adopted a flour consisting of one part wheat, one rye, and one barley, but their labourers would have nothing to do with it—they said they had 'lost their rye teeth', Eden reported in his *State of the Poor*.[2] The government's measures included import bounties on maize and rice as well as grain, farmers were offered premiums to grow potatoes, and parish officers encouraged to give labourers allotments for potatoes and other vegetables. On the whole, however, it is improbable that the government's policies had very much influence on the situation, although the stoppage of the distilleries in 1800 was thought to have saved as much as 360,000 quarters of grain.[3]

Meanwhile, landlords and farmers generally enjoyed highly prosperous conditions. Land was in great demand; the number of Acts of Enclosure was at a peak during these years, and the boundaries of cultivation advanced up the sides of downlands and moors to a point reached neither before nor since. On the loams and lighter soils widespread improvements in cultivation were encouraged, and even the high-cost arable claylands were made very profitable by the wartime price levels.

Following on the rise in farmers' profits came rents. Thomas Tooke believed that during the war period rents rose threefold above the pre-war level. He stated that:

Every purchase of land previous to 1811, whether made with or without judgment, turned out favourably according to the then market rates, and it was supposed, in consequence, that money could in no way be so profitably employed as in buying land. Speculations, therefore, in land, or, as it is termed, land-jobbing, became general, and credit came in aid of capital for that purpose. A striking, but not, I believe, a singular instance of that description of speculation, was exhibited in the case of a petition presented to Parliament some years after, representing that the petitioner had, in the years 1811 and 1812, laid out £150,000, partly his own and partly borrowed, in the purchase of land, which had since fallen so much in value, that he was ruined by the loss; praying, therefore, to be relieved, by what it has been the

---

[1] See W. M. Stern, 'The Bread Crisis in Britain 1795-6' *Economica* (May 1964) pp. 181-6.
[2] F. M. Eden, *State of the Poor* (ed. A. G. L. Rogers, 1929) p. 105.
[3] Cf. Olson, *op. cit.* pp. 56, 70n.

fashion to term an equitable adjustment of contracts, but which means in reality, an indemnification for bad speculations.[1]

Tooke was, of course, justified in drawing attention to the extent of unsound speculation in land brought about by the extraordinary and mainly temporary wartime circumstances, but he exaggerated the size of the increase in rents generally. It is a testimony to the dramatic effect on rentals, however, that the Ricardian theory of rent, based on the increased returns provided by already cultivated land when farming is extended to inferior lands, was formulated in this period. There were indeed some cases of individual farms rising three-fold or even four-fold in value in the war period, but the more general figure seems to have been of the order of about 90 per cent (see fig. 4, page 167). This was the typical increase on large estates whose records have been investigated, and it fits in with the evidence collected by the Board of Agriculture.[2] According to the Board, the expenses of cultivating 100 acres of arable land nearly doubled between 1790 and 1813. The figures obtained by the Board were summarized as follows:

|  | 1790 £ | 1790 s. | 1813 £ | 1813 s. |
|---|---|---|---|---|
| Rent | 88 | 6 | 161 | 13 |
| Tithes | 20 | 14 | 38 | 14 |
| Rates | 17 | 14 | 38 | 19 |
| Wear and Tear | 15 | 13 | 31 | 3 |
| Labour | 85 | 5 | 161 | 13 |
| Seed | 46 | 5 | 98 | 18 |
| Manure | 48 | 3 | 37 | 7 |
| Team | 67 | 5 | 135 | 0 |
| Interest | 22 | 12 | 50 | 6 |
| Taxes | — | — | 18 | 1 |
|  | £411 | 17 | £771 | 14 |

It will be seen that according to the Board of Agriculture's figures agricultural labour was one of the farm expenses which nearly doubled,

[1] Tooke, *op. cit.* I p. 326.

[2] See F. M. L. Thompson, *English Landed Society in the Nineteenth Century* (1963) pp. 217–20; H. G. Hunt, 'Agricultural Rent in South-East England, 1788–1825' *Ag. Hist. Rev.* VII (1959) p. 100.

and there was on average a rise in money wages of about 75 per cent over the whole war period. The big rise in wages was due in part to the rapid extension of cultivation through enclosure, and the associated work of clearing, hedging, ditching, roadmaking and so forth, and partly to the expansion of employment in industry and the armed forces. In areas affected by recruiting shortages of farm labour were severely felt, especially at harvest time, while farmers were encouraged to adopt the newly-invented threshing and winnowing machines designed to economize in barn labour in the winter.[1] Nevertheless, in other areas the increased wages were granted tardily, and as late as 1800 many labourers were still paid only 2s. 6d. more a week than before the war, despite the great increase in food prices. Real wages therefore sometimes lagged badly, and in famine years labourers faced starvation: Arthur Young pointed out in the critical year of 1801 that the quantity of food which a labourer could once have bought for 5s. would now cost him 26s. 6d., supposing he had the money to pay for it. It was little wonder that the Speenhamland system of supplementing wages spread rapidly after 1795, and indeed served to prevent utter destitution and starvation in the years of high prices.

As is well known the Speenhamland 'dole' had numerous defects, although we are now less certain of the seriousness and extent of the defects than we once were. The allowance in aid of wages worked against the single labourer, who was usually not allowed to draw it, and against the married man who tried to keep himself independent of the parish. By tying relief to the size of the family as well as the price of bread it may also have encouraged, as Malthus argued, a tendency towards early marriages and over-large families which was already strong among the poor generally. Together with the Law of Settlement it undoubtedly had the effect of reducing the freedom, independence and self-respect of the pauperized labourer. In the areas affected, the high poor rates fell heavily on the small farmers who employed little or no labour and so did not enjoy the rather dubious benefit of parish-subsidized labour, and this rate burden was perhaps one factor in the small man's decline.[2] And the administration by amateur village officials often led to waste, favouritism and corruption.

But whatever the defects of the wage allowance, it was basically a

[1] See E. L. Jones, 'The Agricultural Labour Market in England 1793–1872' *Econ. Hist. Rev.* 2nd ser. XVII (1964–5) p. 324.
[2] See the critical discussion of the traditional view of the Old Poor Law in M. Blaug, 'The Myth of the Old Poor Law and the Making of the New' *Jour. Econ. Hist.* XXIII (1963) pp. 151–78.

humanitarian policy which helped keep alive a swelling rural proletariat at the expense of farmers' profits and landlords' rents. It might be thought that a system of minimum wages would have been preferable. Indeed, a number of reformers tried without success to press this course on Parliament, but the whole idea ran counter to current economic thought, it posed administrative problems with which the age was ill equipped to deal, and most important, it was in any case an inappropriate solution. In conditions of sharp fluctuations in prices, rapidly increasing population and local surpluses of labour, a minimum wage was, as Pitt realized, impracticable. It would have reduced rather than increased employment, it could not have been easily adapted to meet the needs of both the single man and the man with a family to keep, and it would have left unsolved the problem of the able-bodied unemployed in areas of labour surplus. The Speenhamland system came into existence as a sensible expedient to meet the distress caused by a temporary dearth of corn. But in 1795 it could not be foreseen that prices would continue high for 20 years, and in the event the wage allowance was adopted as a long-term solution of the problem of inadequate wages in those rural areas in which the growth of population, and lack of alternative outlets for labour, created a pool of able-bodied men in excess of local requirements.

Even in the eighteenth century the growth of industry in the Midlands and north influenced the supply of agricultural labour and the wages paid. There developed the division between the semi-industrialized high-wage counties of the northern half of the country and the agricultural low-wage counties of the south. In this period and down to the middle nineteenth century English farm labour was normally mobile only over fairly short distances. The working population of large towns like Manchester and Leeds was recruited predominantly from the neighbouring counties. Consequently, the effects of higher wages and greater variety of employment were felt most strongly near the centres of growing industry, and weakened as distance increased. In the north farm wages were highest near the industrial towns; in the south wages were generally lower, but were lowest in areas remote from London. The Speenhamland system thus developed in the low-wage areas of the south, and was essentially a response to surplus labour and low wages rather than the cause of pauperism among the able-bodied.[1] However, during the war years, when prices were high, farmers were prosperous, and many members of the wealthy classes were urgently seeking ways of helping the poor with allotments, cow-pastures, potato patches and other schemes, the Speen-

[1] Blaug, *op. cit.* p. 169.

hamland allowance gave rise to little criticism. Moreover, in some areas, even in the south, the boom among those industries expanded by the war, and keen recruitment for the armed forces, gave rise to periodical shortages of farm labour. It was only after the war when prices fell, farmers went bankrupt, and landlords were obliged to reduce their rents, that complaints of the poor rates mounted. The costly existence of a permanent surplus of able-bodied farm labourers was now made more evident by the greater real burden of the poor, as the population continued to grow and the decline of many rural handicraft trades threw more hands on to the parish. It was at this point that the ratepayers suspected with more certainty an evil connection between outdoor relief, idleness, large families, unemployment and high poor rates.

A remarkable aspect of the later eighteenth century and the war period was the growth of a sense of solidarity among the farming interest. The country gentlemen and larger farmers, educated and independent by nature, began to draw together to further their particular interests. They belonged to the 'landed interest' in the broad sense of the term, and yet they were independent and critical of it, for the interests of the farmers were not always exactly those of the landlords who made policy in Parliament. Already before the wars the farmers had vented their opinions in Young's *Annals of Agriculture*, and they had vainly contested in Parliament the questions of tithe commutation and the continued prohibition on wool export. Then for a while after 1793 it seemed that the newly-formed Board of Agriculture would become their champion. The Board undertook to make a survey of the farming of the whole country, it acted as a forum for new ideas and the discussion of technical problems, and it pressed for one of the country gentlemen's most favoured reforms, a Bill to cheapen enclosure.

In the result the Board proved a disappointment. Its ill-planned, and frequently ill-executed, county *Reports* lacked utility and failed to sell; it had no facilities and but little money for encouraging technical experiments; and the General Enclosure Bill, after a number of false starts, was finally enacted during the food crisis of 1801 as a dismal half-measure. The combined opposition of the Church and the lawyers, concerned about their tithes and fees, was sufficient to emasculate it, and in the end the Act did little to reduce the expense of enclosure or to simplify the procedure of tithe commutation. The Board was really in an impossibly weak position: financed in a meagre fashion by the government, it yet had no official status and little influence on the government's view of agricultural questions. Its aristocratic oligarchy failed to win the lasting support of the mass of landed gentry and farmers, and after the hollow success of the General Enclosure Act it declined into somnolence and obscurity.[1]

However, a year before Waterloo the fall in prices and the near overthrow of the French Emperor stirred agriculturists, and even the Board of Agriculture, to reconsider the matter of agricultural protection. During the earlier years of the wars protection had not been a very live issue. With wartime prices the Corn Law of 1791 hardly had a chance to operate, and although the sharp fall in prices in 1803–4 had led to a new measure, the Act of 1804, this too was ineffective in the conditions of the subsequent years. But in 1814 the price of wheat once more fell below 75s., and 1815 brought a further fall. The prospect of peace raised the spectre of huge imports from the continent, and it was argued that although continental farmers could produce their wheat at 40s. per quarter, in England it was no longer possible to show a profit at under 80s. Eighty shillings for wheat became the touchstone of future prosperity, county agricultural associations were formed to mobilize rural opinion and bring pressure on Members of Parliament, and the *Farmer's Journal* became the vehicle for the opinions of the extreme protectionists among the 'farming interest'.

Eighty shillings a quarter, a famine price back in the halcyon 1780s, was everywhere put forward as the reasonable level for the prohibition of imports. It was pointed out that during the wars enormous amounts of capital had been devoted to expanding cultivation and improving production, while since 1795 the costs of cultivation had doubled and the landlords' share of the gross proceeds had fallen. The question attracted much attention, of course, and the more disinterested advocates of protection, such as the Reverend T. R. Malthus, lent support with impressive and much broader arguments. In a pamphlet of 1815 Malthus argued that protection was necessary in order to keep farming profitable, and he pointed out that the prosperity of agriculture was not merely a sectional matter but was vital to the well-being of the country as a whole, since the consumption of landlords, farmers and labourers made up a large proportion of the total consumption.[2] This was indeed an important and valid consideration, because at this time agriculture contributed a third or rather more of the national income and gave employment to about the same proportion of the employed population. (A point generally over-

---

[1] See Rosalind Mitchison, 'The Board of Agriculture (1793–1822)' *English Historical Review* LXXIV (1959).

[2] Barnes, *op. cit.* pp. 130–1.

looked, however, was that a large minority of farmers were not engaged in arable production at all and, indeed, had an interest in keeping corn cheap; these, however, seem not to have been vocal or influential.)

David Ricardo, another leading economist and the friendly antagonist of Malthus, gave the most influential and best-reasoned exposition of the free-trade case. While Malthus emphasized consumption, Ricardo accentuated the importance of production, and particularly the role in it of capital. In his view there was no justification for the protection of capital invested in the cultivation of inferior lands—landowners and farmers, just as much as industrialists, should expect to lose capital super-seded by new inventions or more efficient areas of production. Ricardo was not afraid that dependence on foreign grain would be dangerous in time of war, for he had been much impressed by the great expansion of home production which high wartime prices had brought about and could presumably bring about again; and he also believed that if foreign nations were encouraged by free trade to expand their production for the English market the corn trade would become so vital to their economies as to make it impossible for them to let it be cut off in time of war.[1]

In 1815, however, the legislators were not yet ready for free trade in corn. Agriculture still ranked as the greatest industry and the greatest employer; the farmers were vocal in the press and on the platform, and the landlords dominated Parliament; and while industry yet remained heavily protected from foreign competition, why should agriculture not be also? The Parliament of 1815, therefore, in spite of numerous petitions from the large towns and the angry London mobs who manhandled Members and broke the windows of the protectionist newspapers, passed the new Corn Law. The Act of 1815 allowed foreign corn to be imported and warehoused duty free at all times, but wheat could be sold only when the home price reached 80s. a quarter, and barley and oats when they reached 40s. and 27s. respectively. (Lower price limits applied to grain imported from the British North American colonies.)

Thus began a new phase in the history of the Corn Laws. A policy of absolute prohibition of import so long as the home price was below the specified level was substituted for the old policy of three rates of duty which applied at three critical price levels. And on the export side the former restrictions when home prices were unduly high, and the provision of bounties when the home price was low, were both abandoned. This was partly in order to give Irish landowners not only profitable access to the

[1] Barnes *op. cit.* pp. 132–3.

English market, but also the opportunity of capturing the West Indies market, the grain exports of Ireland being fairly substantial at this time. Indeed, the pressure for higher protection and free exports had originated in a Select Committee of 1813, under the chairmanship of Sir Henry Parnell, an Irish landlord who had the interests of his homeland very much at heart.[1] As we have seen, the shift of the Corn Laws in favour of the producer was not entirely new. Both the laws of 1791 and 1804 had raised considerably the limits for low duty imports, but the protective bias of these laws had been somewhat masked by their inoperation during the chronic wartime shortages.

The change in the Corn Laws involved a change in the position of the landlords. As Barnes has put it, the new Act 'left the landlords posing as the protectors of the tenants and agricultural labourers, and facing the hostility of the city labourers, annuitants and manufacturers ... they succeeded, almost as a matter of course, to the position of hatred and opprobrium which the corn dealers and millers had occupied for so many centuries in the eyes of the common people.'[2] Byron's lines expressed the public wrath at the landlords' rapacity:

> *Safe in their barns, these Sabine tillers sent*
> *Their brethren out to battle—why? for rent!*
> *Year after Year they voted cent. per cent.,*
> *Blood, sweat and tear-wrung millions,—why? for rent!*
> *They roar'd, they dined, they drank, they swore they meant*
> *To die for England—why then live?—for rent!*
> *The peace has made one general malcontent*
> *Of these high market patriots; war was rent!*
> *Their love of country, millions all misspent,*
> *How reconcile? by reconciling rent!*
> *And will they not repay the treasures lent?*
> *No: down with everything, and up with rent!*
> *Their good, ill, health, wealth, joy, or discontent,*
> *Being, end, aim, religion—rent, rent, rent!*[3]

The principle of a single price level, below which import was prohibited and above which there was unrestricted import, was soon seen to work badly. Poor harvests in 1817 and 1818 sent prices up to well above 85s., but at first imports were slow to respond since supplies were also short

[1] Barnes, *op. cit.* pp. 117, 121, 142.
[2] *Ibid.* pp. 148, 151.
[3] *Ibid.* p. 184.

on the continent. Then in 1818 when the continental position improved large imports flooded into England. The home price sank continuously between 1818 and 1822 from 86s. 3d. to 44s. 7d., the effects of the large imports of 1818 being reinforced by exceptionally favourable harvests in 1821 and 1822. Gradually it dawned on landlords and farmers that the Corn Laws would not keep prices below famine levels if continental harvests were as poor as the English ones, nor would the total prohibition of import when corn was below 80s. keep prices at a profitable level in England when nature was bountiful. In practice, the sudden turning on and off of the import tap as prices fluctuated above and below 80s. added to the instability of the grain market.

When prices fell to unusually low levels, as in 1815–16 and particularly in 1821 to 1823, there were cries of 'depression' and pleas for help to the legislature. On the first occasion the government reluctantly responded by repealing the income tax and war malt duty, which with other remissions involved a loss to the revenue of some £11,000,000. The total tax revenue in 1816 amounted to only £56,000,000 while the payments on the national debt and the Sinking Fund alone came to £44,000,000. The government was therefore obliged to resort to heavier indirect taxation and to borrowing, and the relief of the property owners and farmers fell heavily on the poor consumers of such goods as soap, butter, cheese, spirits, coffee and tea. In the early 1820s, however, the government would make no further large reductions in the taxes paid by the landed interest: it argued that the agricultural distress was exaggerated and in any event could not be cured by fiscal measures, and so the agriculturists had to be satisfied with a new Corn Law. The Act of 1822 made no practical difference, however, because its graduated scales of duty did not begin to operate until wheat reached 80s, and in fact this was never again to last for longer than a few months during the rest of the nineteenth century. The new law thus still provided absolute protection until a further Act in 1828 abandoned altogether the principle of one fixed price level and returned to the old practice of a range of prices and duties, now embodied in an elaborately graduated sliding scale.

The whole Corn Law question, however, was of little relevance to the problem of the arable farmers. With an acreage greatly expanded by wartime enclosure, and still slowly expanding, and with output threatened by the spread of the four-course rotation and by the return of more favourable seasons, they were unable after 1819 to obtain even in poor years prices much above 65s., while in plentiful years prices sank below 50s., even in some years below 40s. The average price of wheat between 1820 and the repeal of the Corn Laws was 57s. 9d. It was useless for the landed interest to blame the weight of taxation or the deflationary effects of the return to gold in 1821. Landlords and farmers had to accept the fact that the post-war price level was not going to be very much higher than that of pre-war, and readjust themselves accordingly. Wheat imports continued to grow rapidly: the average of the 1820s was 816,000 quarters; that of the 1830s 1,468,000 quarters. But imports were growing less rapidly than consumption, and in the 1840s the proportion of the population fed on foreign wheat was lower than in the 1800s. Furthermore, landlords and farmers could take comfort from the 1826 *Report* of William Jacob, which showed that the English market could not be swamped by European grain. Contrary to the arguments of protectionists, Jacob said, no very large surpluses existed in Europe, nor were they likely to come into existence. The English corn producers could rest assured that very nearly the whole of the home market would remain theirs.

In the post-war years there was much talk of 'agricultural depression'. Periodically as prices fell unusually low a chorus of complaint arose and the government re-examined the Corn Laws and appointed Select Committees to enquire into agricultural distress. Upon examining the evidence the Committees and the government came to the conclusion that little could be done, that the depression was at most regional and intermittent, and that beyond relieving land of some of the burden of taxes, rates and tithes the salvation of landlords and farmers lay in their own hands. Agriculture now had to face the long-term consequences of the intensive application of capital to both old-cultivated and marginal soils. Brougham summed up the problem in a debate of 1816:

Not only have wastes disappeared for miles and miles, giving place to houses, fences and crops; not only have even the most inconsiderable commons, the very village greens, and the little strips of sward along the wayside, been in many places subjected to division and exclusive ownership, cut up into cornfields in the rage for farming; not only have stubborn soils been forced to bear crops by mere weight of metal, by sinking money in the earth, as it has been called—but the land that formerly grew something has been fatigued with labour, and loaded with capital, until it yielded much more; the work both of men and cattle has been economized, new skill has been applied, and a more dexterous combination of different kinds of husbandry been practised, until, without at all comprehending the waste lands wholly added to the productive territory of the island, it may be safely said, not perhaps that two blades of grass now grow where one only grew before, but I am sure that five grow where four used to be.[1]

Brougham was right in emphasizing the increased productivity of agriculture as well as the extension of cultivation to former waste lands. The average yield of wheat was rising rapidly in the first half of the nineteenth century, rising by perhaps as much as a third between the end of the wars and the Repeal of the Corn Laws. There was heavier stocking of land and consequently more manure, and more attention was given to drainage and other aspects of cultivation. With heavier stocking of arable farms and earlier fattening arising from the expanding supplies of rich and more varied fodder and the development of breeds marked by a propensity to early maturing, so also the supplies of fatstock to the market greatly increased.[2] Between the end of the eighteenth century and the 1840s agricultural output was tending to match or even overtake the growth of consumption, but the market effects of this were obscured at first by the abnormal weather, inflation and importing conditions of the war period. When these wartime conditions disappeared prices were bound to fall, and then the farmers with the heaviest costs—notably those on the cold wet clays—were bound to complain.

Strictly, there was no depression in the sense of a long, almost unrelieved, period of unprofitably low corn prices, as in the fourth and fifth decades of the eighteenth century. Between 1815 and 1846 the average prices of wheat, barley and oats all ruled above the pre-war levels, and only in four years out of the 32 did the price of wheat fall below 50s., which in fact had been near the average in the later eighteenth century. When meat prices settled down in the years after 1825 they, too, averaged rather higher than the pre-war level (see fig. 2). Meat had been very dear from 1808 to 1813. Then prices fell sharply by over a third, and although there was a short-lived recovery between 1818 and 1820, they had fallen again in the early 1820s before coming to a degree of equilibrium at about 3s. 6d. to 4s. per stone.

Thus after the initial post-war fall in prices, depression due to low prices was severe and widespread only in two brief periods, 1821 to 1823, when fatstock producers and dairymen were affected as well as grain producers; and in 1833 to 1836, when complaints were confined mainly to arable farms, particularly those on the undrained clays. This is not to say, of course, that other years of the period presented no difficulties for

[1] Barnes, *op. cit.* p. 160.
[2] See R. M. Hartwell, 'The Standard of Living during the Industrial Revolution' *Econ. Hist. Rev.* 2nd ser. XVI (1963–4) pp. 144–5.

farmers. Generally, the low-price years of good weather and high yields of crops and livestock were fairly brief interruptions in a series of bad or indifferent seasons when the advantage of higher prices was offset for many farmers by poor harvests, dear fodder and outbreaks of the rot. 1816, for instance, was a disastrous year with a wet, late and poor harvest and severe losses among the lambs. 1818, by contrast, saw a severe drought with thin corn crops and reduced supplies of feed, giving rise to early sales and low prices for lean stock. 1821, 1823 and 1824 had heavy rains in summer and autumn and severe outbreaks of sheep rot. 1825 and 1826 were years of prolonged summer drought with consequent fodder shortages in the succeeding winters. 1828 to 1830 produced three wet summers in a row with poor harvests, and in 1830-1 appeared the most serious outbreaks of sheep rot of the whole century apart from those of 1879-80. Then in the years 1832 to 1836 a return of better weather conditions brought low prices for both corn and livestock and renewed murmurs of 'depression'.[1]

Farmers seem to have been most vocal in the low-price years. In the November of 1821 when William Cobbett was in Hampshire he heard many complaints from the farmers. 'I have seen a farmer here who can get (or could a few days ago) 28s. a round for a lot of Fat South-Down wethers, which cost him just that money, when they were lambs, *two years ago!* It is impossible that they can have cost him less than 24s. each during the two years, having to be fed on turnips or hay in winter, and to be fattened on good grass . . . fat hogs are sold at Salisbury at from 5s. to 4s. 6d. the *score* pounds, dead weight. Cheese has come down in the same proportion. A correspondent informs me that one hundred and fifty Welsh sheep were, on the 18th of October, offered for 4s. 6d. a head and they went away unsold! . . . The following I take from the *Tyne Mercury* of the 30th of October. "Last week, at Northampton Fair, Mr Thomas Cooper, of Bow, purchased three milch cows and forty sheep, for £18 16s. 6d.!" The skins, four years ago, would have sold for more than the money . . . nothing can be clearer than that the present race of farmers, generally speaking, must be swept away by bankruptcy, if they do not, in time, make their bow, and retire.'[2]

To many farmers, indeed, the outlook seemed so gloomy at this time that there developed a considerable interest in emigration. Some of those burdened by high rents, tithes, and taxes, found overwhelmingly attractive the prospect of becoming the unfettered owners of cheaply-bought

[1] Jones, *op. cit.* pp. 160–5.
[2] W. Cobbett, *Rural Rides* (Everyman edition, 1912) I pp. 13–14.

farmland in America; and they sold up and sailed away to try their fortunes by the shores of Lake Ontario or on the prairies of Illinois, Indiana and Wisconsin. Farm labourers also turned their faces westwards, and the activity of emigration societies, the appearance of books recounting the travels of British farmers in America, and the publication in newspapers of letters from established emigrants, all indicated the deep *malaise* which existed among the British farming community.[1]

The farmer's complaints could be justified when their costs did not fall as rapidly as did their prices. Rents had about doubled during the wars, but it should be remembered that a part of this rise represented the return on landlords' investment in enclosure and other permanent improvements. At first landlords were reluctant to make firm reductions in their rentals. Temporary abatements were freely given, however, in order to placate the farmers and keep them in occupation, and in the distress of 1815 and 1816 the Board of Agriculture's enquiries showed that many landlords had rents unpaid and farms falling into hand, despite abatements averaging about 20 or 25 per cent. In the early 1820s some landlords, but not all, were obliged to make considerable permanent reductions in rentals; and by the mid-1830s, while some landlords maintained their rents at wartime levels, others had reduced them by up to 25 per cent. There was no widespread rise again until the later 1830s but it seems that farmers in general were then paying rents which were still not far from double the pre-war level, while their prices had fallen to a level only 10 to 20 per cent higher than pre-war. It should be noticed, however, that many landlords now thought it wise to step up very considerably their outlay on repairs and improvements, so accepting greater responsibility for the provision and maintenance of the fixed capital of the farms than they had done during the wartime years of high prices. Their tenants were thus relieved of substantial outlays. In doing this, the landlords were helped by lower peacetime taxes. Consequently, the government unintentionally came to the rescue of the farmers by the release to them, via the landlords, of resources from the public revenue.[2]

The wages of agricultural labour fell fast and far immediately after the war in response to the demobilization of the forces and the post-war unemployment in industry. The general level dropped by about a third from 12s. to 15s. a week in 1814 to 9s. to 10s. in 1822, and they rose only a little during the following two decades. Wages, therefore, fell very much

[1] See W. S. Shepperson, *British Emigration to North America* (Oxford, 1957) Ch. 2.

[2] F. M. L. Thompson, *op. cit.* pp. 235–7.

in line with prices, but with the revival of industry and an expanded demand for labour created by the still-growing cultivated area and the increase in yields, they levelled out at about three-quarters of the wartime peak of farm wages, or some 40 per cent above the pre-war level. Taxes, too, became less burdensome with the abolition of the income tax and reductions in the malt tax and duty on horses. The fall in wages, however, tended to be compensated by high poor rates, which pressed particularly hard in the corn counties of the south and east, and little could be done about this, or about the tithes nuisance, until the legislation of the middle 1830s.

In the final analysis much depended on the enterprise and efficiency of the individual farmer and the assistance given him by his landlord. The resources of both might be severely strained where the farms consisted of wet, cold clays or recently-enclosed commons and waste. Some of the common and waste land taken in during the height of the wartime enthusiasm was too thin and impoverished to be worth cultivating in post-war conditions. The heavy clays, too, usually ill drained, and always expensive and difficult to cultivate, became marginal when prices ruled below 60s. The six ploughings and five harrowings of heavy land, the limited range of crops that could be grown, and the low average yields— little more than two-thirds of the national average—made its cultivation arduous and unrewarding. Much of the now marginal land was allowed to tumble down to grass, and on the clays the farmers reverted to the old system of two crops and a fallow, if indeed they had ever departed from it. A Bedfordshire farmer, William Bennett, summed up the position before the Select Committee of 1836: since the big fall in grain prices, he stated, there was no system of management by which the heavy lands could be made to yield a profit; and those farmers who had found their working capital reduced by losses had no option but to stick to the old methods. It was significant that few complaints came from the pasture areas, where the growth of urban markets was expanding the demand for meat and dairy produce, nor from the more cheaply worked and more profitable light soils. Farmers on the free-draining uplands suffered less from the severe outbreaks of sheep rot and were in a better position to reap advantage when mutton and wool prices rose, and Norfolk, said a witness before the Committee of 1833, was 'in as good a state as ever I knew it'.[1]

Recovery was associated with more efficient and more appropriate farming. It is often not appreciated how much agricultural development stemmed from the stimulus of low prices, bad seasons and the threat of

[1] Clapham, *op. cit.* I pp. 133-5.

bankruptcy. The connection between rising prices and the great extension of cultivation and improvements between about 1760 and the end of the wars is well established and widely recognized. But it needs also to be pointed out that low prices could have similar effects. In this way the low-price years of the first half of the eighteenth century had much in common with the difficult years after Waterloo: in both periods there were adjustments in land-use, stocking and rotations to take the best advantage of the markets, and there followed improvements in farm buildings and other changes designed to achieve more efficient farm units. The difference between improvements in periods of low prices and those in periods of prosperity was really one of emphasis. Both low and high prices resulted in a search for greater efficiency: but in the first the emphasis was on greater economy through reduction of costs; in the second it was more concerned with expansion of the cultivated acreage and higher output.

When it was appreciated that at the post-war price levels only efficient farming could be made to pay, there was renewed interest in techniques and improvements and a willingness to invest in them. The progressive landlords and farmers drained and provided new buildings, stocked the light soils more heavily, applied 'artificials' and used machinery to economize in labour. On the heavy lands things continued much as before, but the arable acreage was reduced, and the burdens of poor rates and tithes declined. Prices played a part. Barley and wool prices rose, meat and grains generally were beginning to edge upwards, and for twelve years after 1836 the annual average price of wheat never fell below 50s.

One consequence of the post-war depression was a further decline in the numbers of small occupying-owners. Their ranks had been considerably thinned during the years of difficult farming conditions in the first half of the eighteenth century, when much of their property that came on the market went to swell the estates of the great landlords. With the growing agricultural prosperity after 1750, however, the decline of owner-occupiers had been checked, and indeed there may well have been some increase in their numbers. This at least is what the land tax returns suggest, and some evidence in contemporary writings, and the genial nature of economic conditions of the time lend support to the view. When prices fell after the war many small owners must have found the times unusually difficult. During the years of prosperity some of them had borrowed to buy land, rebuild their farmhouses, find dowries for daughters or place younger sons in a trade or profession; others still carried mortgages originating in enclosure costs and the expenses of restocking their farms. When their profits fell after the war it might be difficult to sustain these debts, and the financial difficulties associated with the return to gold in 1821 and the consequent reduction in note circulation forced borrowers to call in their loans. Some owner-occupiers were working with borrowed capital and were hit by the widespread failure of country banks after the war and the resulting tightness of credit. Others found it more profitable to sell their land and use the capital to rent and stock a larger farm, or to go into some new line of business altogether. And in some parts of the south the poor rates had risen to the monstrous height of 20s. or more per acre—as much as the land was worth in rent—and this sometimes proved an impossible burden when prices fell.

For all these reasons numbers of small owners sold out in the 1820s, and when their land was not bought by the gentry and owners of large estates it was acquired by the growing class of successful professional men, prosperous merchants, country bankers and retired city men—Cobbett's 'tax-eaters' and 'fund-lords'. Nevertheless, the decline in numbers of small owners was rather less marked than the shrinkage of the area they occupied. There was perhaps more of a tendency for the larger of them to sell, while others stayed on by selling off odd acres to meet debts and keep their heads above water. It is impossible to measure the change except by the roughest of guesses. Towards the end of the nineteenth century small owners (excluding the country gentlemen and large owners with home farms and land in hand) occupied about 12 per cent of the cultivated acreage.[1] It seems probable that at the height of their wartime prosperity they had about 20 per cent of a rather smaller acreage. The decline was considerable but not cataclysmic.

One complication in the calculation is that small *owners* were not necessarily equally small farmers: a proportion of them—how large it is hard to say—rented acreages as big or much bigger than the area they owned. The general trend was still towards larger farms, although such a trend necessarily worked slowly. By renting additional acres some small owners were able to farm more efficiently and so survive. In 1851 it was still true that over 60 per cent of all farmers with more than five acres had less than 100 acres and could be classified as small farmers or smallholders. But such occupiers had in their hands quite a small proportion of the acreage in England and Wales—less than 22 per cent—while large farms of 300 acres and over accounted for more than a third of the total acreage. Judged by acreage, the medium-size farm of between 100 and 300

[1] Clapham, *op. cit.* II p. 261.

acres was the typical farm unit and dominated the countryside. Of course, it was true also that the genuinely small farmers could make a reasonable living where conditions enabled them to concentrate on dairying, market gardening, fruit, hops and poultry. Such specialization had long been profitable round London, and the rapid growth of other centres of large consumption, boasting a swelling middle class and even large numbers of comfortably-off skilled artisans, must have considerably expanded the market for this type of farming.

In nineteenth-century England small farms and smallholdings remained everywhere a common feature of the agrarian structure. If they occupied little more than a fifth of the land, the small occupiers yet formed a bridge, an independent class, between the great capitalist farmers and the rural proletariat. The census of 1831 showed that in the whole of Britain there were 686,000 families of agricultural labourers, a proportion of five labouring families to every two families who occupied land. As Clapham said in commenting on the significance of these figures: 'They are entirely destructive of the view that, as the result of agrarian change and class legislation, an army of labourers toiled for a relatively small capitalist class.'[1]
However, nearly a half of the occupying families employed no hired hands regularly, so that the larger farmers employed an average of nearly five labouring families apiece, and if Scotland were omitted from the calculation, the figure would rise to a little over five. The labouring families were distributed fairly generally over the countryside, but tended to be concentrated in the areas of large farms in the south and east of the country. Arable farming was generally carried on in moderately large units, in some areas indeed in farms of over 500, even 1,000 acres, and arable gave rise to most of the regular full-time employment as well as much additional work in harvest time. The western area, where grass predominated, was more typically the home of the family farmers, and there the labouring families often had to supplement farm employment with work in the woodlands, in rural industry, and the carrying trade. The difference should not be exaggerated, for there were after all plenty of pasture farms of 100 to 300 acres which provided regular employment, while in the arable districts there were numerous small farms and even many smallholdings. But it is important to notice that the returns in connection with the Poor Law inquiries of 1832 to 1834 and the Report on Women and Children's Employment of 1867 indicate that while arable and mixed farming required one man for every 25 or 30 acres

[1] Clapham, *op. cit.* I p. 113.

farmed, for pasture the proportion fell to only one man to every 50 or 60 acres, so that arable and mixed farming with its ploughing, harrowing and harvesting, its hoeing of roots and cultivation of artificial grasses, created much higher labour demands than did permanent pasture. For labouring families, however, the most significant point was probably not the nature of the farming but the availability of alternative and supplementary occupations, and the effect on agricultural wages of a wider labour market; and a greater choice of work was certainly a feature of the more industrialized northern half of the country.

Cobbett found 'the more purely a corn country, the more miserable the labourers', and there was some truth in his assertion. Certainly there was a marked connection between wheat growing, heavy seasonal unemployment in winter, and relief of able-bodied men by the Speenhamland system.[1] Larger farmers were less willing to see labourers spend their strength on their own allotments or potato patches, and the lack of alternative employment in southern arable areas greatly reduced the labourers' independence. In the south and east labourers relied mainly on farm employment—hence their hostility to the threshing machine which reduced their winter work—and they had little or no land of their own, few commons from which to gather fuel and food for the pigs, and they suffered in some areas from the competition in the labour market of displaced rural handicraft workers. Furthermore, as Eden had pointed out, the southern labourer had a greater dependence on shops and middlemen for his flour, bacon, milk and cheese, as well as for his clothes.

Despite the heavy enclosure of commons and wastes, however, and the opposition of the large farmers, the majority of labourers in the south had cottage gardens or access to allotments. The allotment should not be too big or 'you make him a little gardener instead of a labourer', and the rent charged was often at least as much as the farmer would pay for the ground. But a quarter-acre of good land would yield 20 cwt. of potatoes, enough for a family and a few pigs as well, and some labourers added sixpence a week to their incomes by selling their pigs.[2] Usually, cottage gardens and allotments were approved of as useful in keeping labourers out of the beer-house and off the poor rate. Some parish and private allotment owners, indeed, made independence of the parish a condition of tenancy. Acts were passed to encourage the provision of allotments, a

[1] Blaug, *op. cit.* p. 171.
[2] R. Molland, 'Agriculture *c.*1793–*c.*1870' *V.C.H. Wiltshire* IV (ed. E. Crittall, 1959) p. 83.

Select Committee of 1843 pronounced them good both for labourers and for landowners, and there was much technical discussion as to what size of ground, what type of soil, and what produce were the most desirable. In Sussex Mrs Mary Ann Gilbert, the wife of Davies Gilbert, the last President of the old Board of Agriculture, demonstrated that pauperism could be diminished if the unemployed were provided with ground on which to grow their own food. She put paupers to reclaiming and cultivating waste land near Beachy Head, and sent Lord Liverpool some specimens of the potatoes grown 'on the beach'. By 1835 she had 213 allotment tenants and she provided them with much good advice on the use of liquid manure, forking the soil, and stall-feeding of cows, while her two agricultural schools offered the children of the allotment holders an opportunity of learning the rudiments of cultivation as well as the three Rs.[1] The provision of allotments was not the complete answer to rural poverty, however, and after 1845, when the Inclosure Act provided for the setting aside of land as allotments in future enclosures of waste and commons, the experience of the Inclosure Commissioners showed that the local demand for them was not always very strong.

Cobbett measured cottagers' felicity partly by the number of pigs and partly by the dress of the womenfolk working in the fields. But he also had an eye for cottage gardens. There existed, he said, 'in Kent, Sussex, Surrey, and Hampshire, and, indeed, in almost every part of England, that most interesting of objects, that which is such an honour to England, and that which distinguishes it from all the rest of the world, namely, those *neatly kept and productive little gardens round the labourers' houses*, which are seldom unornamented with more or less of flowers. We have only to look at these to know what sort of people English labourers are: these gardens are the answer to the *Malthuses* and the *Scarletts*. Shut your mouths, you Scotch economists; cease bawling, Mr Brougham, and you Edinburgh Reviewers, till *you* can show us something, not *like*, but approaching towards a likeness of *this*.'[2]

The cottage dwellings of parts of southern England were at this time generally well built and commodious as compared with those of the north or Wales, and certainly better than the majority of rural dwellings on the continent. The better cottages in the south had two or more bedrooms, were brick-built with tile or slate roofs and brick or tiled floors. In the north they had been built recently, or at least within the last 50 or 60 years. Much of the new building was carried out by the great owners.

[1] A. C. Todd, 'An Answer to Poverty in Sussex' *Ag. Hist. Rev.* IV (1956) pp. 45–51.
[2] Cobbett, *op. cit.* I p. 87.

The seventh Duke of Bedford, for example, built 288 cottages on his Devonshire estate, and 374 on his Bedfordshire estate.[1] But there were also numerous old one-room mud-and-stud erections, without any sort of privy, or wattle-and-daub houses with a ladder to the sleeping-loft, and still to be seen on the remaining stretches of common land were the turf huts of squatters. In Gloucestershire, for instance, both kinds of cottages could be seen, the old style with one room and a pantry downstairs, and only one or two small rooms above, which with a quarter-acre of ground let for 30s. to 50s. a year; and the new and much larger cottages built by the big proprietors, with sitting-room, pantry and wash-house below, and two or three bedrooms above—dearer accommodation certainly, but a cottage which for the first time made possible a decent home life for farm workers.[2]

The typical cottages of Wales and the north were more nearly of the mud-and-stud or wattle-and-daub description, although here and there more substantial ones of brick and stone prevailed, especially where industrial development and labour shortages encouraged such building. Where enclosure had stimulated change and large farms had been created out of the waste, both the farmhouses and the cottages were roomy and well built, their recent origin sometimes evidenced by such titles as Waterlooville or St Helena Farm. In small farm areas the farmhouses were often barely distinguishable from the cottages, and both might be old and cramped unless some large and progressive owner, like Sir James Graham at Netherby, had set about clearing them away and rebuilding.

After the post-war fall in wages had ended in 1824, most labourers still earned substantially more in money terms than they had before the wars. It is true that prices did not fall quite to the pre-war level and taxes were heavier, but there can be little doubt that the majority of labourers were better off than they had been in 1790. For the next 25 years there was no very large increase in money earnings, nor probably in real earnings either, until in the 1840s there were considerable reductions in the prices of foodstuffs consumed by the poor—some of the first benefits of free trade. There was in fact an increase in money wages in the later 1820s and again in the later 1830s, but a decline set in during the 1840s. In money terms labourers were receiving rather less than 10 per cent more in 1850 than they had in 1795, but the fall in prices raised real wages substantially in the seven years before 1850.

There was a tendency, however, for the gap between the high-wage

[1] D. Spring, *The English Landed Estate in the Nineteenth Century* (1963) p. 52.
[2] *J.R.A.S.E.* XI (1850) p. 177.

counties of the north and the low-wage counties of the south to widen. Conditions in the southern counties remote from London were indeed often desperate. The badly-paid labourers of southern Wiltshire, to take an extreme example, earned a winter rate of between 6s. and 7s. a week in 1794, 8s. in 1804, and 12s. by 1814. But in 1817 they were back to 7s. to 8s. and between 1817 and 1844 their winter rate was not generally much above 7s. Wages were considerably higher in summer, but then the hours were enormously long, in harvest time stretching from three in the morning to eight in the evening. Carters and shepherds received 2s. a week more than the ordinary labourers, and women working in the fields got three-fifths of the male labourer's wage. In 1843 the incomes of the labourers of southern Wiltshire ranged between 8s. and 18s.[1]

In Lincolnshire, a county where the labourers were considered 'pretty comfortable', the average wages were about 11s. a week in 1851, and the villages contained numerous friendly societies, savings banks and coal and clothing clubs. In the Isle of Axholme on the north of the county, women and children above the age of ten were regularly employed, and could earn 2s. a day in harvest, 1s. 4d. for taking up potatoes, and 9d. for weeding and planting and sorting potatoes. In other parts of the county, however, wages for women and children were lower, and many families lived in mud-and-stud cottages, all sleeping together in one room.[2] Both Wiltshire and Lincolnshire were primarily agricultural counties with few industrial occupations to attract the labourer away from the farm and pull up the level of wages. A good example of a semi-industrialized and therefore high-wage county is Nottinghamshire with its numerous framework-knitting and coal-mining villages. Here farm wages in the early 1830s were up considerably on the pre-war level and were tending to rise. In 1833 the average weekly wage was about 12s. 6d., while in 1795 it had been 9s. and in 1824 10s. 3d. The total family income might range from 12s. to £1 a week, according to whether the wife and children were out at work. There was evidence of improvement in living standards, with the diet including meat as well as bacon, and wheaten bread and milk as well as the vegetables provided by the garden or allotment, while coal was easily come by. Friendly societies were spreading in the villages, and sometimes a labourer had savings up to a sum equal to two or three times his annual wages.[3]

[1] Molland, *loc. cit.* pp. 80–1.
[2] *J.R.A.S.E.* XII (1851) pp. 404–8.
[3] J. D. Marshall, 'Nottinghamshire Labourers in the Early Nineteenth Century' *Trans. Thoroton Soc.* LXIV (1960) pp. 60–4.

It is clear that the number and ages of the children made a great difference to a family's income and comfort. Some figures for labouring families in Norfolk and Suffolk for 1839 show that a family with four children over the age of ten earned altogether twice as much as a single man (£51 a year compared with £25), and over half as much again as a family whose children were all under ten. Each child over ten added an average of £4 3s. to the family's annual income.[1] But too often the children's precious supplement was earned in the degrading and demoralizing gang labour, and this was especially the case in the eastern counties, in areas where land had been recently enclosed and adult labour was scarce. Of course, the good years when the family's income was high were offset by the more numerous years when the children were too young to earn or had grown up and left home. According to the Norfolk and Suffolk figures a third of the families had all their children under the age of ten, and the annual income of some 70 per cent of the families was under £36 a year. Throughout the countryside women and children could find some employment, but not so easily in most counties, perhaps, as in Gloucestershire where, we are told, the only farm work in which they were not employed was ploughing, mowing, and hedging and ditching. For everything else—planting and weeding, hoeing and reaping, and in winter cutting up turnips for feed, turning the chaff-cutting machines and assisting with threshing—women and often children too were in demand.[2]

It was poverty and the longing for a taste of meat, rather than a mere disrespect for property or want of diversion, which turned labourers into poachers and made the woods ring with nocturnal alarms. The Game Laws were not new, but in the early nineteenth century they reached a new level of severity, and in some years two-thirds or more of all convictions were those under the Game Code. It was significant that the offences were greatest in the south where the labourers were poorest, and were most frequent in the years of greatest distress.[3] For a time the warfare between gamekeepers and poachers was a savage affair of spring-guns and buckshot, but the spring-gun was prohibited, and in 1831 the sale of game was legalized, leading to a decline in the organized poaching gangs. But there still remained the village lads who for the sake of a few rabbits or hares, pheasant, trout or salmon were prepared to risk imprisonment and transportation and the chance of being maimed for life.

In their well-known work, *The Village Labourer*, the Hammonds dwelt

[1] H. Fearn, 'Chartism in Suffolk' in *Chartist Studies* (ed. A. Briggs, 1959) p. 152.
[2] *J.R.A.S.E.* XI (1850) pp. 163–4.
[3] F. M. L. Thompson, *op. cit.* p. 143.

at length on the Game Laws, the severity of the sentences, the common occurrence of poaching, and the isolation felt by the poor as commons disappeared and landlords, parsons and farmers erected stronger barriers against any trespass on their private property. 'The woods in which Tom Jones fought his great fight with Thwackum and Blifil to cover the flight of Molly Seagrim now echoed on a still and moonless night with the din of a different sort of battle: the noise of gunshots and blows from bludgeons, and broken curses from men who knew that, if they were taken, they would never see the English dawn rise over their homes again: a battle which ended perhaps in the death or wounding of a keeper or poacher, and the hanging or transportation of some of the favourite Don Quixotes of the village.'[1]

Writing of a later and less violent period, W. H. Hudson told many anecdotes of the life on Salisbury Plain—the stratagems designed to deceive the gamekeepers, the pitched battles in the woods, and the hiding in the cottages to avoid arrest. At the end of the century the old hostility and isolation of the labourers still continued: 'The keeper is there to tell him to keep to the road and sometimes to ask him, even when he is on the road, what he is looking over the hedge for. He slinks obediently away; he is only a poor labourer with his living to get, and he cannot afford to offend the man who stands between him and the lord and the lord's tenant. And he is inarticulate; but the insolence and injustice rankle in his heart, for he is not altogether a helot in the soul; and the result is that the sedition-mongers, the Socialists, the furious denouncers of all landlords, who are now quartering the country, and whose vans I meet in the remotest villages, are listened to, and their words—wild and whirling words they may be—are sinking into the hearts of the agricultural labourers of the new generation.'[2]

The lowness of wages and insufficiency of winter employment, together with limited mobility and the increase in the rural population, explained the continuing high poor rates and prevalence of pauperism in the southern agricultural counties. In the south of Wiltshire, admittedly an area of exceptionally low wages and extremely poor conditions, there were many parishes where there was not a single labourer's family not drawing something from the poor rates. Because of poor relief and the farmers' practice of turning off the single men first when winter came, youths of 17 commonly wed girls of 15, and bred up large families. As one writer has commented, 'Ireland was not the only place in the British Isles

<hr/>

[1] Hammond, *op. cit.* pp. 162–75.
[2] W. H. Hudson, *A Shepherd's Life* (14th ed. 1933) p. 77.

where the potato prevented a population from starving.' It is not surprising that those removed from want felt compelled to provide gifts of meat and corn, coal, blankets and clothing, nor that emigration, sometimes organized, sometimes voluntary, was common.[1] Caird's description of the diet of the typical labourer on Salisbury Plain is worth giving:

We were curious to know how the money was economised, and heard from a labourer the following account of a day's diet. After doing up his horses he takes breakfast, which is made of flour with a little butter, and water 'from the tea-kettle' poured over it. He takes with him to the field a piece of bread and (if he has not a young family, and can afford it) cheese to eat at mid-day. He returns home in the afternoon to a few potatoes, and possibly a little bacon, though only those who are better off can afford this. The supper very commonly consists of bread and water. The appearance of the labourers showed, as might be expected from such meagre diet, a want of that vigour and activity which mark the well-fed ploughman of the northern and midland counties. Beer is given by the master in hay-time and harvest. Some farmers allow ground for planting potatoes to their labourers, and carry home their fuel—which on the downs, where there is no wood, is a very expensive article in a labourer's family.

Both farmers and labourers suffer in this locality from the present over-supply of labour. The farmer is compelled to employ more men than his present mode of operations require, and, to save himself, he pays them a lower rate of wages than is sufficient to give that amount of physical power which is necessary for the performance of a fair day's work. His labour is, therefore, really more costly than where sufficient wages are paid; and, accordingly, in all cases where task-work is done, the rates are higher here than in other counties in which the general condition of the labourer is better. We found a prevalent desire for emigration among the labourers themselves, as their only mode of benefitting those who go and those who remain behind.[2]

At the middle of the century wages remained miserably low in the southern counties generally—below 9s. a week on average in 1850—where the lack of alternative occupations was sorely felt and where the proportion of labourers to the cultivated acreage was high. The influence on wages of alternative employment could be readily seen even in some areas of the southern counties, as in Somerset, where the highest farm

[1] Molland, *op. cit.* pp. 83–4.
[2] Sir James Caird, *English Agriculture in 1850–51* (1851) pp. 84–5.

wages were paid in the area near the coal pits.[1] In the northern high-wage counties, the area according to Caird north of a line running through the middle of Shropshire and Leicestershire to the Wash, where agricultural wages felt the stimulus of industrial pay rates and the attraction of the towns, they averaged as much as 3s. (or 37 per cent) higher than in the south (see his map, fig. 3). The range of wages ran from the peak of 14s. in the industrial West Riding to as low as 7s. in the unspoiled but indigent countryside of Gloucestershire, Wiltshire and Suffolk.[2]

The most recent opinion discounts the traditional view that the pauperizing of able-bodied labour in the southern counties resulted from the operation of the infamous Speenhamland system. All the evidence suggests that low wages and dependence on relief was the result of the large stagnant pool of surplus labour and the accompanying high structural and seasonal unemployment. Heavy poor law expenditure was the consequence of low wages and unemployment, and not the cause. But some counties, even in the south, had never operated the Speenhamland system; and recent study of the problem shows that by 1834, when the Speenhamland allowance in aid of wages was abolished by the New Poor Law, the giving of such allowances had very much declined, probably as a result of the severe criticism and enquiries into the system which had developed since the wars, but probably owing something also to improvement in the labourer's position. Allowances to married men with more than three or four children were still widely given, however, in the low-wage areas. It is not only the effects and extent of the Speenhamland system that have been exaggerated, but also the number of the recipients. By 1834, it is now thought, able-bodied labourers and their dependants in receipt of relief probably did not exceed 300,000.[3]

Under the Old Poor Law, therefore, the labourer enjoyed some degree at least of social security, although too often the relief was provided in a form that was degrading and humiliating. The more serious defects of the poor law system resulted not from the practice of giving outdoor relief but from the operation of the Settlement Law. In particular, the Settlement Law tended to confine the labourer to his parish and restricted by artificial means his naturally limited mobility, and it often forced him to live in an overcrowded and high-rented cottage at a distance of perhaps several miles from his employment. These effects of the Settlement Law

[1] *J.R.A.S.E.* XI (1850) p. 750.

[2] Caird, *op. cit.* pp. 511–12.

[3] Blaug, *op. cit.* pp. 169–70, 177; see also his 'The Poor Law Report Re-examined' *Jour. Econ. Hist.* XXIV (1964) pp. 231, 241.

3 *Caird's outline map of England, showing (a) the division between the low-rented arable area of eastern England and the high-rented western pasture area, and (b) the division between the northern high-wage area and the southern low-wage area.* (English Agriculture in 1850–51)

were felt in the eighteenth century and persisted throughout the first half of the nineteenth century. It was not until 1862 that the poor law union rather than the individual parish became the unit for settlement purposes, and long before that the interest of landowners and farmers in denying labourers a settlement, and thus a claim to relief, had led to the growth of 'closed' parishes. These were parishes where the proprietors were sufficiently few to act in concert: they pulled down cottages and built no new ones, and of necessity their labourers had to live in an adjoining parish and find a home and their settlement there. The obverse of the 'closed' parish was the 'open' one, where small farmers, village shopkeepers and tradesmen made a profit out of running up cheap, small cottages for the labourers. Louis Simond, a visiting Frenchman, was struck by the apparent comfort of the cottagers and the absence of poverty in the countryside, until at last he encountered in one place a curious scarcity of cottages. He was told that such cottages 'were nests of vermin, pilferers and poachers; far from building, they would rather pull them down'. The labourers 'reside in some small town or village in the neighbourhood, and walk several miles to and from their work'. Commented Simond: 'There are, then, it seems, obscure corners where the poor are swept out of the way.'[1]

An example of the particular local reasons for the creation of these closed and open parishes was described by Caird when he visited Wiltshire in 1850:

> There is much complaint among the farmers of the severity of the rates, which a variety of causes has produced. The village inhabitants of these districts [in northern and western Wiltshire] are principally a decayed manufacturing population, among whom handloom weaving and pillow-lace working still keep a languid existence. A man who is a weaver himself, or descended from weavers, is not held in much estimation as a farm labourer; and in this grazing district, where the demand for labour is not great, the land must bear the burden of a population not required for its cultivation—those manufactures on which they formerly subsisted having ceased to afford them support. The surplus labour is not divided among the farmers by mutual agreement, as is common in other districts, where the tenants are men of capital. The fear of this pressure, aided by the present law of settlement, has induced large proprietors to diminish the cottages on their estates, and thus the burden is increased on those open parishes to which the population is driven. In the union of Melksham some of the parishes

---

[1] W. Smart, *Economic Annals of the Nineteenth Century* (1910) I p. 312.

have no paupers, having cast their labourers off upon their unfortunate neighbours, where property being more divided and cheap, cottages are run up on speculation, and the new comers are welcomed by a certain class whose property is thereby enhanced to the heavy loss of the ratepayers. The tyranny practised on the poor labourer when he falls into arrear to his new landlord is great. This man, often the keeper of a huckster shop where the labourer gets his various wants supplied, charges every article he sells at exorbitant prices, from which there is no appeal, as, if the labourer leaves his residence, he cannot get another.[1]

The demand for cottages in open parishes was reinforced by the gradual decline of the practice of 'living-in', and the boarding of the labourers in the employer's farmhouse. Still strong in the north, the boarding labourers could yet be found in a few of the large farmhouses in all the southern counties, but the changing habits of the wealthy farmers worked against it. It was now '*Squire* Charington and the *Miss* Charingtons; and not plain Master Charington and his son Hodge, and his daughter Betty', exclaimed a peevish Cobbett, when he saw a gentleman farmer's establishment in Surrey. 'There was hardly any *family* in that house where formerly there were, in all probability, from ten to fifteen men, boys and maids: and, which was the worst of all, there was a *parlour*. Aye, and a *carpet* and *bell-pull* too! ... And there were the decanters, the glasses, the "dinner-set" of crockery-ware, and all just in the true stock-jobber style. ... Why do not farmers now *feed* and *lodge* their work-people, as they did formerly? Because they cannot keep them *upon so little* as they give them in wages.'[2]

It was a combination of these circumstances, the low levels of wages and the constant struggle to exist, the game laws, the degradation of the poor law, the decay of living-in and the growth of rural slums, the immobility of surplus labour and the lack of alternative occupations, together with the loss of winter employment to the threshing machine, which led up to the labourers' revolt of 1830. To Cobbett, moving round the countryside in the spring of that year, the signs of approaching revolt were very evident. Near Lincoln he saw

three poor fellows digging stone for the roads, who told me that they never had anything but bread to eat, and water to wash it down. One of them was a widower with three children; and his pay was eighteenpence a day; that is to say, about three pounds of bread a day each, for six days in the week:

1 Caird, *op. cit.* pp. 75–6.
2 Cobbett, *op. cit.* I pp. 265–6.

nothing for Sunday, and nothing for lodging, washing, clothing, candle-light, or fuel! Just such was the state of things in France at the eve of the Revolution! Precisely such; and precisely were the causes.

And near Leicester he was struck by the contrast between the beautiful churches, the impressive vicarages, and the 'miserable sheds' of the labourers:

> Look at these hovels, made of mud and of straw; bits of glass, or of old cast-off windows, without frames or hinges frequently, but merely stuck in the mud wall. Enter them, and look at the bits of chairs or stools; the wretched boards tacked together to serve for a table; the floor of pebble, broken brick, or of the bare ground; look at the thing called a bed; and survey the rags on the backs of the wretched inhabitants; and then wonder if you can that the gaols and dungeons and treadmills increase, and that a standing army and barracks become the favourite establishments of England![1]

In the summer of that year the men's grievances welled up into a violent but selective anger against the symbols of their misery—the barns, ricks and poor-houses which they burned, the overseers whom they evicted from their villages in the indignity of the parish cart, and the hated threshing machines which they sought out and smashed. Beginning in Kent, and generally taking the form of a determined demand for another 3d. or 6d. a day, the movement spread westwards as far as Dorset, Wiltshire and Gloucestershire, and northwards to East Anglia and Northamptonshire. The main strength of the risings, however, lay in the southernmost counties. Some magistrates, and even many of the farmers, sympathized with the labourers and recognized the justice of their claim for a living wage. At first, therefore, arrests were few and higher wages were conceded. By December, however, the magistrates, encouraged by the government and the support of larger bodies of troops, felt strong enough to take the offensive and put down outbreaks as soon as they appeared. The revolt collapsed once opposition stiffened, and the rioters paid a heavy penalty for their show of force: six were executed, over 400 transported, and about the same number imprisoned at home. Thus ended the last revolt of the English agricultural labourers.[2]

The failure of the outbreaks of 1830, and the harshness of the retribution exacted by the law, could not crush entirely the labourers' spirit. Some

---

[1] *Ibid.* II pp. 253-4, 266.
[2] See the vivid account in J. L. and B. Hammond, *The Village Labourer* (1912), Ch. XI-XII. Somewhat similar factors were important in the 'Rebecca Riots' in Wales in the 1840s.

refused to face a life that offered no prospect but a constant struggle to find work and the bare essentials of existence, with the poorhouse as the refuge at the end of it. The more enterprising sought jobs in the towns, or emigrated and encouraged others to follow, as the Petworth labourers did. As early as 1833, a Petworth Emigration Society published a shilling booklet of letters from former Sussex men who had gone to Canada, and the demand for the booklet was so heavy that a second edition was printed within the year.[1] Yet the slow reduction of the surplus agricultural labour force in the south bears witness to the very real obstacles to mobility—the lack of money and the tendency towards early marriage and the inescapable burden of a young family; the continued operation of the settlement law and its effect of tying the pauper to his parish; and the fear of the unknown and untried, a deep ignorance of any kind of life more than 10 or 20 miles away, when in southern England the nearest industrial town might be London, and 50 miles seemed as far as 500. Furthermore, the labourers' betters, his squire, farmer and parson, were not disposed to help or enlighten him. Their concern was to keep labour obedient and cheap, and if they encouraged emigration it was only as a means of lightening the poor rates.

Indeed, the growth of numbers and the poor rates, and the consequent degradation of the labourers—first by the roundsman system or weekly auction of pauper labour, then work on the roads or parish farm, harnessed perhaps to a cart, or the final indignity of the harsh post-1834 work-house—this was the wedge which prised apart the old rural society. Justices, farmers and parsons had a common interest in keeping down the swelling labour force which threatened to engulf their rents, profits and tithes, to say nothing of the threat to law and order, and morality, in the countryside. Thus arose the concern with the poor laws, the rigid enforcement of the game code, and the granting of allotments and premiums to those labourers who kept their families off the parish. The labourer felt alienated and socially isolated by the hostile front of restriction and disapproval presented to him by the propertied classes. And if the great landowner stood detached above these village politics, he would not, or could not, do more than keep some kind of balance between the rival interests in the village.

The labourers, finding no help where they might traditionally have expected it, turned to self-help. In the general enthusiasm for unionism of the early 1830s there were even attempts by the agricultural workers to force wages up by combination. The story of the 'Tolpuddle martyrs'

[1] Shepperson, *op. cit.* p. 10.

is too well known to bear retelling here. But the vicious fate of that small band of Dorsetshire labourers, whose attempt to raise wages ended disastrously in an unwitting breach of the law, showed that times were not ripe for agricultural unionism, as they were not to be for many a long year ahead.

The 100 years after 1750 had brought in turn prosperity and difficulties for landlords, small owners and farmers. But if there were difficult times there were also good ones, and there can be little doubt but that the owners and farmers of 1850 lived far better than did their predecessors of three generations back. To a degree this was also true of a large proportion of the labourers, and particularly those in the high-wage counties of the Midlands and north. But it was not true of all. If the labourers of the low-wage areas had made any progress it was mainly in their housing conditions and the growing possibility of a better life in the towns. It can hardly be doubted that these small gains were insufficient to outweigh for many a certain loss of independence and a deep long-prevailing sense of injustice which continued to rankle until the nineteenth century itself was a forgotten thing of the past.

*Suggestions for further reading:*

| | |
|---|---|
| D. G. Barnes | *A History of the English Corn Laws* (1930), Ch. VII–VIII. |
| Sir James Caird | *English Agriculture in 1850–51* (2nd ed. 1966). |
| J. H. Clapham | *An Economic History of Modern Britain*, I (2nd ed. 1930), Ch. IV, XI. |
| W. Cobbett | *Rural Rides* (Everyman edition, 1912). |
| C. R. Fay | *The Corn Laws and Social England* (Cambridge, 1932). |
| J. L. and B. Hammond | *The Village Labourer* (1912), Ch. VIII–XII. |
| E. Hobsbawm and G. Rudé | *Captain Swing* (1969). |
| R. Mitchison | 'The Old Board of Agriculture (1793–1822)' *English Historical Review* LXXIV (1959). |
| W. S. Shepperson | *British Emigration to North America: Projects and Opinions in the Early Victorian Period* (Oxford, 1957), Ch. 1 and 2. |

# 6 The Corn Laws and the Landed Interest

As we saw in the previous chapter, the object of the Act of 1815 and of the subsequent Corn Laws was to secure profitable prices for English grain producers by maintaining a high level of duties on imports. In this the Corn Laws failed, for the main factors in the price of English grain were on the one hand the domestic harvest and the great expansion of the cultivated acreage since the later eighteenth century, and on the other hand the rapidly growing population. In so far as they kept out imports, the Corn Laws necessarily tended to raise prices above the level that would have obtained without them, but down to Repeal in 1846 the influence of protection on prices was somewhat marginal. Even when conditions of free entry prevailed, it was only from the 1860s that grain imports amounted to more than a fifth of total consumption; although for wheat alone the proportion was higher—over a third of the wheat consumed was imported.

The Act of 1828 had replaced the fixed level of 80s. at which wheat might enter the country by a sliding scale of duties, the duties falling as the home prices rose. When the average home price was below 52s. absolute protection still continued; but at 52s. foreign wheat was admitted on paying a duty of 34s. 8d.; when the home price rose to 53s. the duty fell to 33s. 8d.; and so on with gradually reduced duties for each shilling rise in the home price; until with the price at 70s. the duty was only 10s. 8d., at 71s. it was 6s. 8d., 72s. 2s. 8d., and finally only a nominal duty of one shilling was levied when the home price reached 73s. or more. A much lower range of preferential duties applied to wheat from the colonies, and similar sliding scales were devised for other grains. In effect, while absolute protection now applied at a much lower price than 80s., the high duties at the top of the scale meant that little foreign corn would normally come in until the home price was in the fairly high range of 60s. to 70s.

Under this much more flexible system grain imports increased considerably, and especially so between 1828 and 1831, when the crops at home were poor and high prices attracted imports of over 2,000,000 quarters of wheat a year. Between 1832 and 1836, however, the harvests were good and prices came down, wheat falling in December 1835 to the very low level of 35s. 4d. Under these conditions there was little real ground for complaint about the Corn Laws. Their opponents accepted that they did not keep imports out when prices at home were high, and the landed interest no longer expected them to keep prices up when harvests were exceptionally favourable. It could still be objected, however, that the Laws gave rise to speculation, and that corn merchants deliberately rigged the market, delaying shipments from arriving until the home price should have risen and the import duty payable correspondingly reduced. But rather than tinker further with this unsatisfactory system of protection, landlords and farmers turned to the reform of the Poor Law in 1834 and the commutation of tithes in 1836, with the hope of lessening agriculture's burdens and creating greater prosperity.

Before the late 1830s circumstances determined that the demand for free trade was heard only from the Benthamite Radicals, who had studied their Ricardo and were convinced of the validity of the economists' 'natural laws', and by some thoughtful Whig landowners, like Earl Fitzwilliam, who had come to the conclusion that the Corn Laws helped the landlords much less than they believed, while definitely harming the farmers and labourers, and through dear wages, the manufacturing interests. Outside the textile districts there was at this time, however, little general interest in free trade. The landed interest as a whole were of course opposed to it, and so were the majority of manufacturers and merchants, long accustomed to a protected home market and a regulated colonial trade. Indeed, even the limited reductions in duties and freeing of trade which Huskisson had achieved in 1824 to 1826 aroused strong protests from the commercial interests affected, and until the 1840s free trade made no further progress, for the few reductions which were made in some duties were counterbalanced by increases in others. In these years the interest of Parliament was absorbed in Catholic emancipation and religious questions, in the momentous struggle for parliamentary reform, the Factory Acts, education, municipal corporations, and the Poor Law, and there was little desire or opportunity to reconsider the question of free trade.

Moreover, as the small group of ardent free traders recognized, there was one very important obstacle in the way of any large reductions in tariffs, and that was the practical question of finding alternative revenue. In 1830 import duties produced 43 per cent of the government's annual revenue, and import duties and excise together accounted for as much as 75 per cent of the total. Clearly, if there was to be any real measure of free trade some large alternative source of revenue had to be found. In practice this could only mean reviving the income tax (which at the time of its abolition in 1815 had produced 22 per cent of the revenue), for in 1830, while the taxes other than customs and excise produced only a quarter of the revenue, they generated a disproportionate quantity of resentment and political heat. It was unthinkable that these taxes could be stepped up to any considerable extent: the free traders were really in the situation where their objective not only failed to command general support but was not even practical politics, unless and until some turn of events brought into power a government willing and able to take the highly unpopular step of reintroducing the income tax.

There was little hope after the election of 1837 that the weak Whig government would be prepared to envisage any radical changes in fiscal policy. Yet it was just at this time that the free trade tide began to run more strongly. The free trade radicals had argued that high food prices, fostered by the Corn Laws, not only injured the working classes but also had adverse effects on trade and industry. Dear food reduced the purchasing power of the population for manufactured goods, while our failure to buy the food exports of our foreign customers prevented them from buying as much of our manufactures as they might, and stimulated them to develop their own rival manufacturing industries. Protection was blamed for the stagnation of trade in the 1830s; and slow trade was held to mean low profits, which spelled an unfavourably slow growth of industrial resources. Even some of the great landlords were impressed by these arguments. Agriculture was the base of English prosperity, Earl Fitzwilliam once wrote, but 'manufacture was the shaft and commerce the capital of that column, and if the shaft and capital were destroyed the base would be useless.'[1]

In the late 1830s the change in economic conditions gave fresh point to

[1] D. Spring, 'Earl Fitzwilliam and the Corn Laws' *American Historical Review* LIX (1953–4) p. 291.

the free traders' views. Bad harvests led to high prices and unprecedentedly heavy corn imports—between 2,000,000 and over 3,000,000 quarters of wheat a year between 1839 and 1842—and to a correspondingly heavy drain of gold abroad to pay for them. The banking system was weakened, and a terrible depression in the cotton industry created famine conditions—the whole town of Stockport was said to be 'to let', and families in Nelson and Colne sold all their furniture for food and lived on milk and boiled nettles. Protection, the supposed cause of the depression, was now widely attacked, and in 1840 the government's Select Committee on Import Duties produced its famous *Report*.

The Committee concluded that the chaos of import duties was greatly in need of simplification, and that many duties could be safely abolished without affecting the revenue because, in fact, almost all of the customs revenue was produced by only 17 articles, of which corn and wool were two. On some of the 17 articles it would be possible actually to increase the yield of revenue by reductions in the rates of duty. And in regard to the Corn Laws, the Committee accepted the opinion of the Board of Trade officials that foreign tariffs were raised in retaliation to our own, and that duties on imported grain raised the level of wages in Britain and therefore made our export prices less competitive. At the same time, and rather contradictorily, it was held that repeal of the Corn Laws would pose no immediate threat to English agriculture because even in the years of bad harvests and high prices experience had shown that imports were small in relation to the home production; while in any case, it was argued, a free market in grain might be healthier for English farmers, leading them to concentrate less on arable and to turn some of their land to profitable alternative uses for which it was better suited.

The *Report* was well received by the press and public, and indeed was regarded as an authoritative statement of the rights and wrongs of the free trade controversy—this in spite of the fact that the evidence was heavily weighted by free trade opinion and was of dubious validity, and that the Board of Trade officials had been unable to provide firm support for their contention that British exports were in fact affected by our tariffs. The Committee itself was packed with free traders, and it even contrived to hold its meetings in the summer months when it was known that its few representatives from the landed interest were likely to be out of London visiting their estates.[1]

The rise in food prices and the industrial depression after 1838,

[1] See the account in Lucy Brown, *The Board of Trade and the Free Trade Movement, 1830–47* (1958) pp. 70–5, 141–213.

together with the *Report* of the Select Committee on Import Duties, were all excellent ammunition for the heavy propaganda artillery of the Anti-Corn Law League. The appearance in 1838 of this organization, dedicated to the one object of achieving free trade in corn, was a new factor in the history of the Corn Laws. The League was formed at an auspicious point in that history, when a combination of dear food and unprecedented industrial distress provided the conditions under which it was possible to create a large body of opinion favourable to reform. The place of the League's foundation, and its permanent headquarters, Manchester, was symbolical of the League's strength. The cotton industry was not only the great example of the triumph of machinery in industrial production, and of the rising power of industrial capital, but it was also the leading export industry, and the one most vitally interested in the free trade issue.

Ostensibly, the Anti-Corn Law League based its appeal on the economic advantages of free trade—cheaper food, more employment, higher exports and greater prosperity. Repeal was depicted as a great humanitarian measure which would reduce poverty and overcome pauperism by the influence of better conditions on the labouring classes. (The League played down the possible effect of free trade in bringing about a reduction in wages, however, for it was afraid of antagonizing the working classes.) Once free trade was universally adopted it would usher in peace between nations, for no country would go to war knowing that it depended on other powers for its essential supplies. Free trade thus touched such fundamental issues that it took on almost the character of a religious crusade. Indeed, dissenting clergymen—the attack on the landed interest was incidentally an attack on the Church of England—preached sermons against the Corn Laws, and Cobden claimed that protection was 'opposed to the law of God', and was 'anti-scriptural and anti-religious'.[1]

But while the Leaguers spread their economic principles and invoked for them the authority of divine blessing, at bottom their concern was with political consequences. Repeal of the Corn Laws would have beneficial economic results certainly, but its political results would be even more desirable to the radicals who made up the hard core of the movement and for whom the League served as a rallying point—nothing less than a fatal weakening of the landed interest and the overthrow of the Tory party. The grand objective of the League was to adjust the political power in the country in accordance with the change that had occurred in the economy,

[1] See N. McCord, *The Anti-Corn Law League* (1958) pp. 103–7; G. Kitson Clark, 'The Repeal of the Corn Laws and the Politics of the Forties' *Econ. Hist. Rev.* 2nd ser. IV (1951–2) p. 5.

to take the governmental power of the landed interest and put it in the hands of the merchants and manufacturers. Bright made this clear when he described the League as a movement 'of the commercial and industrial classes against the lords and great proprietors of the soil'.[1]

How was this to be done when landowners made up the majority of both parties' Members of Parliament? The Reform of 1832 had enfranchised many industrial towns and had increased the number of members whose affiliations were with trade and industry, a number which in fact had been growing for half a century before 1832. But the situation remained that the landowners still commanded two-thirds of the seats in the Commons, and were only slightly less numerous among the Whigs than among the Tories. On the other hand, there was and had long existed a division in the ranks of the landed interest, not merely over the political and religious principles which divided Whig and Tory, but over agricultural policy itself. The division was between those landlords who were inclined to consider agricultural questions in the context of the economy and the interest of the nation as a whole, and who were prepared to think that it was in the long-run interest of landowners that the national interest should prevail over the purely agricultural one, and those other landowners who narrowly regarded agriculture as all-important to the nation and themselves, and protection as vital to this all-important sector of the economy.

The League recognized the existence of this division among the landed interest, and appreciated, too, that even among the small country gentlemen and farmers opinion was divided on the Corn Law question, for those who were mainly concerned with livestock and produced or bought grain as fodder had an interest in keeping it cheap, while those who catered for industrial markets, like the dairymen of Cheshire and Lancashire, wanted above all good trade. The success of the League depended on its ability to exploit these divisions, and its hope was that if free trade could be made a major national issue commanding widespread support, it could be the lever which would find and prise apart the fissures in the façade of landed unity.

To this end the League set out to mobilize public opinion, the first of the modern propaganda machines and one of the most strikingly successful. At first, it is true, its campaigns were hampered by lack of funds and by public apathy, but gradually, with the financial support of the cotton manufacturers and merchants, they achieved results. The printed word of free-trade journals and pamphlets was supplemented by the oratory of

[1] Kitson Clark, *loc. cit.* p. 5.

pulpit and platform. Parcels of free-trade tracts arrived on the doorsteps of electors, and working-class lecturers toured the towns and countryside to deliver free addresses on the subject of the 'Accursed Corn Laws, in demonstration of their blasphemous and unnatural character, their in-human tendency, and their destructive consequences to every interest in this vast community—especially to the industrious agriculturists, and not even excepting that of the lord of the soil, the legislative monopolist, the feudal inflictor of universal calamity . . .'. The League never minced its words, and variously described the landlords as 'a bread-taxing oligarchy, a handful of swindlers, rapacious harpies, labour-plunderers, monsters of impiety, putrid and sensual *banditti*, titled felons, rich robbers, and blood-sucking vampires'; while their dupes the farmers were more contemptu-ously named 'brute drudges, clodpates, bullfrogs, chawbacons and clodpoles'. The League's representatives, it it not surprising to find, were not everywhere well received. In some villages they were lucky to escape a wetting in the duck-pond, and at Cambridge the undergraduates—'a gang of unfledged ruffians in cap and gown' according to the League's report—assailed the free-trade lecturer, and 'after exhausting their ob-scene and blasphemous vocabulary, exhibited themselves in the character of prize-fighters with the rest of the audience! . . .'.[1]

To increase its pressure on Parliament, the League also interested itself in the registers of electors, encouraged its supporters to buy the necessary property qualification for the vote, and put up its own candidates to fight elections, so that soon the thunder of Cobden and Bright resounded in the Commons itself. Belatedly, a combination of landowners, clergy and large tenant-farmers, first brought together by Robert Baker of Writtle in Essex, organized themselves in 1844 into an 'Anti-League' to counter the free trade agitation. Local groups of country gentlemen and farmers amalgamated into a national association, and attracted the support of some large owners such as the Duke of Richmond, and found publicity in the columns of the *Farmer's Magazine*. Although in imitation of the League they circulated tracts, employed lecturers and intervened in the registration of electors, they did so in a much less whole-hearted manner, for in general the supporters of the Anti-League disapproved of their rivals' methods. The League's appeal to public opinion was seen by many of the conservative country gentlemen and farmers as constituting a dangerous precedent and a real threat to the constitution, which even after 1832 was still based on a very small electorate. The Anti-League was

[1] Barnes, *History of the Corn Laws* p. 257; McCord, *op. cit.* pp. 56–64, 134–5, 143–8, 180–6.

therefore too cautiously hostile to the principle of political democracy to really oppose the League with its own weapons, and its impact on the Corn Laws issue was limited. Some of the more aggressive figures among the farmers, it is true, were prepared to take the fight into the enemy's camp: 'Why should the landlords have the power to destroy the tenants?' asked Robert Baker. 'If the landlord chose to cut his throat, his tenantry were not bound to do the same', answered his followers. But the main effect of the Anti-League was to make the opinion of agriculturists known to Parliament, and to help strengthen the opposition of protectionist Members within the Tory party in the Commons.[1]

The League itself was so apparently successful that it is easy to exaggerate its influence. The League certainly aroused support for Repeal and made it a national issue which the government was obliged to face, but it must be remembered that it never overcame the apathy and hostility of much of the country. It was essentially a middle-class movement, concerned with wresting for the middle class some of the power of the aristocracy. The educated members of the working classes—the Chartist leaders, for instance—recognized it as such, and regarded with cynicism the supposedly humanitarian character of the League's objective. Free trade might mean cheaper bread, said the intelligent artisan, but if so, might it not also mean lower wages, and a flood of displaced agricultural labourers competing for jobs and houses in the towns? The League could not avoid being associated with a sectional interest—with Manchester and all that Manchester stood for—which it tried to persuade the country was really the national interest. In London and the south the League never counted for very much, and while Cobden, Bright and Villiers brought the free trade arguments into the very heart of Parliament, the landed interest was still as well entrenched there as ever. The Repeal of the Corn Laws was caught up in the tide of events—in the industrial depression of 1839 to 1842 and the great famine in Ireland; it was caught up too in the hazards of politics and the accident of personalities—particularly in the personality of Peel—and if we were to list the factors which led to Repeal the League would rank only as one among many.

The depression of the early forties gave rise to shortfalls in government revenue and so to a large budget deficit, a development which presaged the collapse of the Whig government. In 1841 Peel and the Tories came into office with a strong majority, and with a country willing to

[1] See G. L. Mosse, 'The Anti-League: 1844–1846' *Econ. Hist. Rev.* XVII (1947) pp. 131–42; Mary Lawson-Tancred, 'The Anti-League and the Corn Law Crisis of 1846' *Historical Journal* III, 2 (1960) pp. 162–83.

accept new and firm measures in order to see prosperity restored and the threat of civil disorders, and possibly revolution, forestalled. Peel's immediate task was to put the national finances into order, and to achieve this he took the bold step of reintroducing the income tax after a lapse of nearly 30 years, and of using the additional revenue which the tax provided to experiment with reductions in import duties. These reductions included the duties on corn, for the new Corn Law of 1842 lowered greatly the protection offered by the sliding scale of 1828. There is no evidence that at this stage Peel had been won over to the free trade camp, or that he had long-range plans beyond the three-year term for which the income tax was granted. But he recognized that the country's situation was one which required new policies, and his approach was essentially an empirical and experimental one.

His measures of course succeeded. The budget was balanced and the reductions in duties on raw materials and manufactured goods seemed to be beneficial, for aided by better harvests, cheaper food and heavy investment in railway building, the economy rose out of the slough of depression. For a time the Anti-Corn Law League found conditions working against them and their thunder stolen by Peel's success. By 1845 an emboldened Peel was prepared to extend the income tax for a further three years in order to continue the abolition and lowering of duties. However, his measures, important as they were, remained tentative and exploratory. The whole of his tariff reductions between 1842 and 1846 cost the exchequer only a little over £6,000,000 in revenue, and reduced the proportion of customs duty in import prices only from 31 to 25 per cent.

Nevertheless, before long his policy would inevitably be stopped short by what remained the great foundation stone of the protection system, and the League's agitation would compel him to reconsider the Corn Laws. Had events worked differently, it seems likely that Peel would have preferred a low, perhaps almost nominal, scale of duties on corn imports, rather than have conceded the total repeal which the League demanded. In any case, the gradual reduction of protection for industry sooner or later would have made it morally and politically impossible to exempt agriculture from the rigours of free competition. But events unexpectedly brought the issue to a head. The famine in Ireland and the Malthusian fatalism of the government's Irish policy, the bad harvest of 1845 at home, and the renewed pressure of the League combined to force Peel's hand. By the end of 1845 he had become convinced that a major change in the Corn Laws was immediately necessary.

The famine in Ireland could be met, and was met, by the temporary suspension of the Corn Laws, but Peel was obliged to decide in favour of *total repeal* (within a period of three years) by wider considerations. Like many of his party he saw the attack on the Corn Laws as a constitutional issue, an attempt to mobilize public opinion and override the sovereign power of Parliament. For this reason, it has been argued, he was determined that the change in the Laws should be made by the existing Parliament and not after a dissolution and a general election fought on the issue. Suspension of the Corn Laws was only a temporary expedient and would leave the ultimate issue unsettled; therefore, it had to be repeal, and since he wished to avoid an appeal to the country, the repeal had to be carried through immediately. The proposal to spread the disappearance of the duties over three years helped to mollify agricultural opinion, for it gave farmers some opportunity of adjusting themselves to the new situation, and eased the passage of the Bill through Parliament.

In 1846, therefore, after it had been shown that no other statesman could form a government, Peel introduced a Bill to abolish within three years the protective duties on corn, split his party between Peelites and protectionists, and retired from office. It was a courageous policy, and one which he believed to be in the interests of the country at large. Even the reactionary Duke of Wellington was obliged to agree that good government was more important than the Corn Laws. The result of Repeal, Peel told his Tamworth constituents in 1847, 'tended to fortify the established institutions of this country, to inspire confidence in the equity and benevolence of the legislature, to maintain the just authority of an hereditary nobility, and to discourage the desire for democratic change in the Constitution of the House of Commons'.[1]

Repeal was thus a strategic retreat, the sacrifice of the bastion of the Corn Laws in order to keep intact the main stronghold of aristocratic power and the limited constitution. Peel perhaps underestimated the short-sightedness of some of his party and the abiding strength of their support for the Corn Laws. He was himself a progressive landowner, and was personally impressed by the possibilities for more efficient and more productive farming. He had become convinced that the Corn Laws were not necessary for agricultural prosperity, and that in any case the prosperity of the nation demanded cheapness and plenty. Wages, he now believed, did not move up and down with the price of bread, and the labouring classes would therefore benefit from Repeal. For agriculture he considered that the loss of protection would be amply compensated

[1] See Betty Kemp, 'Reflections on the Repeal of the Corn Laws' *Victorian Studies* V, 3 (1962) pp. 195–204.

by the growth of the home market with the increase in population and the purchasing power arising from a growing trade.

Many of the large proprietors agreed with him, and a majority of the Members whose seats depended on aristocratic favour voted for Repeal, while it was the country gentlemen and farmers, especially in the eastern arable areas of the country, who took up a narrow, reactionary view and clung to protection. Of course, the large owners were accustomed to the responsibilities of office and hardened to the necessities of politics, and they accepted the unpleasant inevitability of a measure which affected their own private interest so closely. They were also more wealthy and could afford concessions, and they drew a large proportion of their incomes from coal mines and ironworks, from canal and railway shares, urban ground rents and government stock, as well as from agricultural land. Nonetheless, their attitude to Repeal was not determined entirely or even principally by their economic interests, for some of the chief coal owners and urban proprietors were among those who opposed Repeal. Ultimately the division among the landed aristocracy depended upon how each owner believed Repeal to affect the political power of the landed interest in the long run. Those who thought that only a realistic view of the place of land in the mid-nineteenth-century economy would enable the landed interest to retain the respect of the nation and continued acquiescence in their leadership, supported Repeal; while those who believed that the political strength of the landed interest depended on its economic strength, regarded Repeal as the great betrayal.[1]

In the event, the Repeal of the Corn Laws did not prove to be immediately disastrous for the landed interest. Although the proportion of landowners in the Commons gradually shrank, they still accounted for nearly a half of the seats at the end of the nineteenth century. Economically the full effects of Repeal were delayed for 30 years. Between 1848 and 1852 there was a sharp fall in grain prices which would certainly have been rather less severe had there still been protection; and the level of grain imports rose substantially over the next 20 years, and by 1872 to 1874 accounted for over 30 per cent of total consumption (and for wheat nearly a half of the total consumption). Nevertheless, during the 30 years after Repeal the price of wheat remained fairly stable, averaging nearly 53s. a quarter, which was only 5s. a quarter less than the average price achieved in the last 26 years of protection; and after 1858 the price fluctuations were also less violent than hitherto (see fig. 5, page 178). The effect of free trade

[1] See G. Kitson Clark, *op. cit.* and Spring, *loc. cit.* pp. 297–304.

was to bring the world price of wheat up to the British level and to make it more stable, rather than severely to depress the home price. G. R. Porter, one of the Board of Trade's advocates of free trade, was over-sanguine, however, in arguing that 'a very simple calculation would suffice to convince' that a country of large population could never become 'habitually dependent on the soil of other countries for the food of its inhabitants'. The very simple calculation referred to was that at the time of Repeal it would have taken twice the total amount of shipping which then entered British ports to supply the country with grain. But by 1871 the quantity of shipping had much more than doubled, and by 1891 had multiplied over five-fold. 'A measure of dependence was now possible', Clapham commented, 'of which, only a generation earlier, no one had ever dreamed.'[1]

But while corn imports grew and wheat prices remained steady, other agricultural prices rose very considerably in the years after Repeal. Imports of barley and oats increased less markedly than those of wheat, and their prices rose by some 20 per cent between 1848–57 and 1867–77. Meat and dairy produce rose even more, by between 30 and 45 per cent (see fig. 5). Once it had recovered from the short-lived depression immediately after Repeal, agriculture was generally and increasingly prosperous, and entered upon what Lord Ernle impressively called its 'golden age'. Looked at from the point of view of agricultural conditions between the 50's and the late 70's Repeal was a spurious issue, and both the free traders and the protectionists exaggerated its significance for agriculture. Only wheat and wool were seriously prejudiced, and even the arable farmers could benefit from the rising prices for barley, oats, hay and straw, which crops together were by 1850 not far from equal in value to the wheat crop. In any case, arable farmers were turning more to producing feed for stall-fed cattle, while the pasture farmers, who occupied nearly a half of the total acreage, found their prices rising quite steadily and substantially.

Although divided politically by Repeal, the landed interest was not therefore very adversely affected in an economic sense. The home market was expanding, as Peel foresaw, through the growth of numbers and the rise of real incomes, and farming efficiency was improved by the railways and better communications, by cheaper drainage, investment in which was aided by government loans, by heavier spending on fertilizers and buildings, and by more flexible crop rotations and readier adaptation to market trends. Farmers' profits rose, and landlords' rents increased by a quarter between 1851–2 and 1878–9. We must leave a more detailed

[1] J. H. Clapham, *The Economic Development of France and Germany* (4th ed. 1936) p. 209.

examination of this period of prosperity for the next chapter, however, and consider now the further attacks which were launched upon the landed interest.

In a great speech in the Commons in 1845 Cobden addressed this appeal to the landowners:

> You gentlemen of England, the high aristocracy of England, your forefathers led my forefathers; you may lead us again if you choose; but though—longer than any other aristocracy—you have kept your power, while the battle-field and the hunting-field were the tests of manly vigour, you have not done as the noblesse of France or the hidalgos of Madrid have done; you have been Englishmen, not wanting in courage on any call. But this is a new age; the age of social advancement, not of feudal sports; you belong to a mercantile age; you cannot have the advantage of commercial rents and retain your feudal privileges, too. If you identify yourselves with the spirit of the age, you may yet do well; for I tell you that the people of this country look to their aristocracy with a deep-rooted prejudice—an hereditary prejudice, I may call it—in their favour; but your power was never got, and you will not keep it by obstructing that progressive spirit of the age in which you live. If you are found obstructing that progressive spirit which is calculated to knit nations more closely together by commercial intercourse; if you give nothing but opposition to schemes which almost give life and breath to inanimate nature, and which it has been decreed shall go on, then you are not longer a national body.[1]

Cobden believed that in order to maintain their political power and retain the Corn Laws, the landowners were refusing to grant their tenants leases, knowing that insecurity of tenure would oblige the farmers to vote as their landlords directed. From this particular criticism it was easy to develop the more general argument that the English landlord-tenant system was inimical to progress, for agricultural experts had long made the granting of leases a principal instrument of good husbandry. Only farmers with security of tenure, the argument ran, would undertake costly improvements and farm the land to the best advantage; and landlords were deliberately neglecting to make the best of their estates, and neglecting also the welfare of their tenants, by refusing leases for political reasons.

This argument came to be frequently heard in the nineteenth century when the technical advantages of heavier investment in soil improvements, advanced crop rotations and better breeds of animals were more widely

---

[1] Barnes, *op. cit*, pp. 265-6.

known and appreciated, and when the attack on the political power of the landed interest developed with the growing political consciousness of the urban middle class. The Corn Law bastion having been taken, the radicals widened their assault to include two other aspects of English landlordism, estate management and the control of land through strict family settlements. The landlords' management of estates was criticized on the grounds that they failed to support their tenants with adequate capital expenditure on such things as drainage and buildings, and that the running of the estate was subordinated to the needs of aristocratic sport, while detailed control was often placed in the hands of ignorant amateurs rather than expert agents. The strict settlement, of course, was a legal device to ensure that a family's property was handed down intact from one generation to the next, and the radicals were right in saying that this had the effect of making the break-up of large estates unlikely, and thus of restricting the amount of land which came on the market.[1]

Nearly 20 years after the Repeal of the Corn Laws Cobden declared: 'If I were five and twenty . . . I would have a League for free trade in Land just as we had a League for free trade in Corn.' At the time he spoke, criticism of the 'aristocratic monopoly' of the soil bid fair to mount into just such another crusade as the attack on the Corn Laws. But before many more years had passed, agricultural depression, the secret ballot, and extension of the franchise robbed the controversy of much of its point, while the practical difficulties of land reform and of restoring a land-owning peasantry proved to be much more formidable than their advocates had supposed. Moreover, the attack on landed property could never secure the support which had carried the League to victory. To too many of the middle class it raised the broader and dangerous question of the rights of private property in general.[2] Eventually, just before and just after the First World War, their own inclinations and the compulsion of death duties led many landowners to dispose rapidly of property, and by this time the issue of land reform was hardly a live one.[3]

The criticism of the landed interest, of course, was not unfounded. There were undoubtedly landlords who refused to grant leases on political grounds, and some farmers who would have farmed better with greater security of tenure. Caird's survey of English farming in 1850–1 showed that there was much backward husbandry, and many estates where land-

[1] See chapter II, pp. 43–5.
[2] Thompson, *op. cit.* pp. 283–5.
[3] See the bibliographical discussion by O. R. McGregor in his *Introduction* to Lord Ernle, *English Farming Past and Present* (6th ed. 1961) pp. cxxxiv–cxxxvii.

lords had neglected improvements, and as a leading, although not entirely unbiased, authority he strongly criticized the landlords' choice of agents. The aristocratic monopoly of land was also very much a reality, for an official enquiry in 1873 showed that as much as half of the country was owned by only 4,217 persons, each of whom possessed 1,000 acres or more.

Yet there is also no doubt that the large element of truth in the allegations against the landowners, and more important the significance of the charges, were exaggerated by the radicals to meet their political purposes. It is true that with the extension of the franchise and the secret ballot, and even more with heavier estate expenditure and the decline in the relative importance of land in the economy, the political and economic position of the landed interest was gradually whittled away in the years after 1846; but the change was a gradual one and the persistence of landed wealth and of landed domination of Parliament invited renewed attacks on the stewardship of the soil. The consideration that none of the criticisms was new, but went back at least to the later eighteenth century, suggests that they were not provoked by a recent decline in standards of landownership but rather by the hostility generated by the growth in political power of the middle classes.

Viewed in their proper context, the radical criticisms had remarkably little force. As a recent historian of the period has commented: 'It is difficult to see that these strictures were fully justified by the economic facts of the time. . ... The great merit of English agriculture as it was around 1870 was the degree to which it had made commonplaces of the most striking innovations developed over the previous century and a half. It had brought to a higher level of efficiency than had been attained earlier or elsewhere an industry whose chief concerns—products, markets, labour supply and social setting—were still determined by a traditional state of affairs in which rural life predominated and which was rapidly fading in England.'[1] It was the reluctance of this traditional rural life to fade more rapidly before the forces of industrialism, that really lay behind the radical concern with leases, family settlements, 'feudal sport' and aristocratic extravagance.

In fact, the most serious weakness in the English landlord-tenant system was exhibited on those estates where the owners' indebtedness and neglect had caused a running down of the land and farm amenities, and created a vicious circle of bad farming, poor tenants and low rents. Even here, however, the situation was not irreversible. Sooner or later it was

[1] W. Ashworth, *An Economic History of England 1870–1939* (1960) pp. 50–1.

likely that a new owner or trustees would take the property in hand, if only to pay off debts by jacking up the revenue with more able tenants and a higher level of rents. Taken as a whole, there was probably much more landlord's capital sunk in farm improvements in the middle years of the nineteenth century than in any comparable period, and estates owned by the leading statesmen and political figures were among the most progressive. Where improvements were not going forward, it was often because the tenants themselves were conservative and had not asked for them. To encourage better farming some landlords tied rents to prices, and made abatements conditional on farmers spending the corresponding amount on fertilizers. But these measures designed to increase competitive efficiency were often opposed by the farmers.

Detailed studies have also shown that despite the lack of formal professional training, and in many cases of experience in farming, estate agents were generally competent chief executives of the great estates, indeed were the key figures in their development, 'efficient, zealous and versatile'. There was in practice a growth of professional standards among agents of the large estates, the newcomers often receiving a training in one of the offices of the greatest owners, where they were exposed to the complexities of changing market conditions and reorganization, enclosure of wastes, canal and railway building, structural works and drainage, before taking on a controlling position elsewhere.[1] And after all, it was not only in agriculture that one could find examples of inadequate investment, outdated equipment, amateurism, and backward techniques.

It is true that rents were rising substantially between 1850 and 1879, but much of this increase represented a return on new capital investment in drainage and sometimes on buildings also, and such investment does not appear to have obtained more than a very moderate return. After such investment there was often a long delay before rents were raised, and the usual return on drainage works was only about 3 per cent or even less. The capital outlay, at about £5 per acre, was at least as heavy as in the great enclosure period, but the return on drainage, in contrast to the 15 or 20 per cent yielded by enclosure, was even less than could be obtained in many other investments, such as the Funds or the railways. Moreover, drainage often involved a heavier stocking of farms, and therefore compelled the landlord to meet a heavier expenditure on buildings and yards. Taken as a whole, landlords' investment in improvements, if judged by commercial standards, could not and did not pay. In effect the landlords

---

[1] Thompson, *op. cit.* pp. 156–61, 177. See also the detailed discussion of estate administration in Spring, *op. cit.* esp. Ch. I, III–IV.

were subsidizing the farmers by an uneconomic use of capital—an expensive aberration that sprang partly from the strength of the tradition of the landlord's function, and partly from a miscalculation as to the long-term returns of high farming. But landlords were not blind to what was happening, and a few of them became so alarmed by the deterioration in the economic and political value of land that they talked darkly of selling up and taking their capital elsewhere.[1]

Family settlements certainly kept large estates together, but it was not unknown for land to be allowed to run out of settlement, and of course, not all of the land of the large owners was subject to settlements in any case. If few or no very large properties came on the land market, there were still very numerous transactions, and estates of many hundreds of acres changed hands not infrequently. In the years of the heavy attacks on the 'aristocratic monopoly of land' before 1880, there was an estimated turn-over of between 300,000 and 600,000 acres; so the land market can hardly be regarded as inactive.[2] There is no evidence, therefore, that would-be landowners were unable to find land to buy, and indeed after 1879 there was a shortage of purchasers. Even had there been a much greater turn-over of land, it is very doubtful whether the control of a different set of landowners would have had any noticeable effect on the rate of agricultural development.

The farmers' grievances over depredations caused by horse and hound and the carefully-preserved game were real ones, but it should be remembered that many of the farmers joined in the sport themselves, and that land liable to this kind of damage was usually low-rented as a compensation. Much more objectionable aspects of rural sport were the harshly-prosecuted Game Laws, but as a reflection of the wealthy classes' exalted belief in the sanctity of private property even the Game Laws were not unique. They had their parallels among the urban middle classes, even probably among those who like Cobden condemned 'feudal sports'. The urban water companies which shamelessly dispensed unfiltered and cholera-infected water, the slum landlords who successfully set at naught the Public Health Acts, and the industrial employers who evaded Factory Acts and neglected safety precautions—these were certainly more destructive to human happiness than the prosecutors of the Game Laws.

Although some farmers complained of the absence of leases, and it seems probable that greater certainty of tenure and of compensation for

[1] Thompson, op. cit. pp. 242–4, 248–50, 255, 290.
[2] F. M. L. Thompson, 'The Land Market in the Nineteenth Century' Oxford Economic Papers IX (1957) p. 300.

unexhausted improvements would have contributed to better farming, yet it is important not to exaggerate this deficiency of English estate management. There are three relevant aspects of the nineteenth-century lease question to take into consideration. First, it should be recalled that in the eighteenth century the use of leases was largely a regional custom and was connected with the size of the farms and the nature of the husbandry. Where the farms were large the landlords usually employed leases as a safeguard against loss of rent and damage to the property, and in some areas leases were valuable for the regulation of permanent pasture, marling and crop rotations. Leases had never become universal, and in the nineteenth century they still remained the usual practice for large farms, mainly in areas where they were well established in the previous century. Over the whole country they lost ground during the Napoleonic Wars, and the nineteenth-century lack of enthusiasm for leases was due not so much to landlords who wished to keep tenants in political subjection but to the marked fluctuations in rents and prices during and after the Napoleonic Wars. This period taught both landlords and farmers the dangers of being tied to fixed rents and farm practices for long terms of years. On many estates it was not that landlords would not grant leases but that tenants would not accept them; and landlords with experience of leases in the 1820s knew that while they prevented them from asking for more rent when conditions were good, they did not prevent tenants from asking for abatements—and getting them—when conditions were bad.

Secondly, leases were not necessarily a guarantee of improved farming. Obviously farmers would be more inclined to make improvements if they felt more certain of reaping advantage from them, but many other factors also influenced the tenants' investment—the levels of prices, for example, and the ability of the farmer to find the money for improvements. Some leases pretended to keep a close control over the farmer's husbandry, but the most progressive farmers did not want leases of that kind. The best farming was highly flexible and closely geared to the market, varying with changing conditions the proportions of grass to arable, and of white crops to fodder crops, and such could not be practised under a restrictive lease. It is significant that Norfolk, the home of the long and regulatory lease, was falling behind as an area of advanced farming owing to the strict following of the four-course rotation. Even where leases were still important for regulating the husbandry it is doubtful how far they were enforced, or indeed were enforceable.

Lastly, it was true in the eighteenth century, and was still true in the middle nineteenth century, that tenancies at will did not necessarily involve insecurity of tenure. Under the traditional arrangement many thousands of farming families went on for generations, often making improvements, and quite confident in their landlords. Even Caird, a strong advocate of leases, gave examples of this mutual trust between landlord and tenant. The majority of such farms were small, and many were backward and unprofitable, but they were not always so, as their rents show. And Caird, despite his general preference for leases, felt bound to point out instances where leases had proved unsatisfactory to one party or the other, and he noted that in some places farmers were unwilling to accept leases. While farmers were often subjected to some degree of pressure in election years, and this was particularly so with farmers in Ireland and Scotland, in England there is no evidence of any great unwillingness to accept the landlord's judgment in political matters or of strongly-felt grievances over leases, game or land laws. Leasehold farmers as well as annual tenants accepted representation by their landlords as a matter of course, and as natural and fitting. In conclusion, it seems clear that leases were far from being universally valuable in contemporary farming conditions; and when considered in the broader context their importance was small as compared with the growth of markets, the improvement of communications, discovery of new techniques, and availability of capital. For all the criticism, their absence did not prevent English farming from being the most advanced of the age.[1]

The Repeal of the Corn Laws showed that the political power of the landed interest, as the significance of land itself, was in decline in the nineteenth century. But down to 1879 the economic strength of the landed interest was still great. The long process of the accretion of land to the larger estates was still going on, but more slowly than in the eighteenth century; in part it continued to proceed at the expense of the small occupying owners, whose share of the cultivated acreage shrank from about a fifth in 1800 to about an eighth or less by the end of the century. The gentry, meanwhile, although not making large territorial acquisitions, seem to have been holding their own with fair success. Many landowners, of course, did not have to rely on agricultural rents exclusively, but drew a part, even in some cases the larger part, of their income from government and Bank stocks, canal and railway shares, urban ground rents, and the profits and royalties of harbour facilities and mines of various kinds,

[1] For a valuable discussion of the lease question from opposing points of view, see McGregor, *loc. cit.* pp. cxxiv–cxxxiii, and Thompson, *English Landed Society* pp. 203–4, 227–31. See also the commentary in Spring, *op. cit.* pp. 178–82.

quarries, ironworks, brickworks and other ventures. Landowners were no longer important in the introduction and early fostering of industrial activities as they had been in the eighteenth century, but their land and capital was still of importance in this development, and the rapid growth of industry and urban population in the nineteenth century made such investments more lucrative than ever.

Agricultural rents were still by far the main source of landed incomes, however, and determined the financial position and outlook of the majority of the landed interest. After a period of uncertainty and, it is likely, one of declining rents for the majority of owners between the end of the French wars and about 1835, rents definitely moved upwards to reach a peak in 1879 at a level which has since been approached only in very recent years. The total gain in rents between 1815 and 1879 averaged between 25 and 45 per cent, but there were wide variations, and the collapse with the great depression brought the average level in 1900 down to about that of 1815. Landlords really made their permanent gain from the doubling of rents which occurred during the wars of 1793 to 1815, for in the long run it was the 1815 rent, or thereabouts, which proved to be the level that could be held (see fig. 4). The rise in rents of the middle decades of the century was not of much advantage, not only because it could not be maintained after 1879, but also because in many cases the increases had been achieved

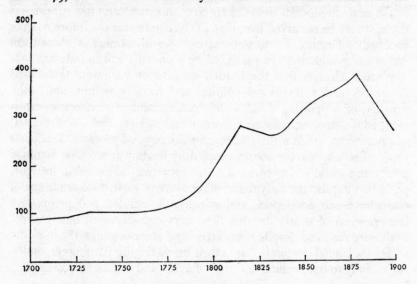

**4** *Rent Movements 1700–1900 (1730–50 = 100)*

only by very heavy capital expenditure on improvements, expenditure which yielded little enough return while rents were still high, and nothing at all when rents fell from 1879. In this last period of rising rents 'the biggest improvers among great landowners were subsidizing agriculture, contributing directly to its over-capitalization, and encouraging further over-capitalization by the tenants who farmed the improved farms'.[1] In all this it should be borne in mind that there was a marked difference between the experience of landlords in the western or mainly pasture half of the country and those in the eastern or mainly arable half. Rents both rose more rapidly before 1879 and fell much less after 1879 in the western half of the country, and rose less and fell more in the eastern half. A considerable difference in average rents per acre was already noted at the time of Caird's survey of 1850–1, and it became even more marked as pasture became increasingly more profitable than arable in the next 50 years (see fig. 3, page 142).

In retrospect much of what came to be known as 'high farming' was a strategic miscalculation, a misdirection of resources. Landlords sank capital in drainage and buildings, and farmers devoted much time and money to developing advanced systems of cultivation, neither of which could pay at the prices which ruled in the last decades of the century and for long after. Exactly how far 'high farming' was justified, we discuss in the next chapter. In so far as an error in agricultural investment was made, it may be remarked here that a failure to foresee the future was not confined to farming, for to some extent over-investment in obsolescent systems of production was paralleled by a similar trend in industry.

What is clear is that the burden on land of supporting a leisured aristocracy and a traditional political and social structure (and, incidentally, of supporting a not so leisured element of aristocratic statesmen, administrators, reformers, writers and military and naval officers) was not incompatible with agricultural efficiency and progress. Trollope's world of sedate country houses peopled by wealthy aristocrats, dignified gentry and worldly clergymen, a world apparently absorbed in intrigue, place-hunting and the daily round of social trivia, existed alongside one of energetic estate developers, and scientifically-minded and progressive farmers—and of nearly destitute labourers—which he all but ignored. Both were real, and despite many errors and short-comings, the justification of the landed interest is that in their hands English farming responded successfully to the swelling demands of a new and urbanized economy.

[1] See F. M. L. Thompson, 'English Great Estates in the Nineteenth-Century, 1790–1914', *First International Conference of Economic History: Contributions* (1960) p. 394.

*References and suggestions for further reading:*

| D. G. Barnes | *A History of the English Corn Laws* (1930), Ch. XI, XII. |
| Lucy Brown | *The Board of Trade and the Free Trade Movement 1830–42* (1958), parts 1 and 3. |
| G. Kitson Clark | 'The Repeal of the Corn Laws and the Politics of the Forties' *Economic History Review* 2nd ser. IV (1951–2). |
| Susan Fairlie | 'The Nineteenth-Century Corn Laws Reconsidered' *Economic History Review* XVIII (1965). |
| C. R. Fay | *The Corn Laws and Social England* (Cambridge, 1932). |
| N. McCord | *The Anti-Corn Law League 1838–46* (1958). |
| D. C. Moore | 'The Corn Laws and High Farming', *Economic History Review* 2nd ser. XVIII (1965). |
| D. C. Moore | 'The Corn Laws and High Farming' *Economic History Review* XVIII (1965). |
| Donald Read | *Cobden and Bright* (1967). |
| D. Spring | 'The English Landed Estate in the Age of Coal and Iron, 1830–1880' *Journal of Economic History*, XI (1951). |
| D. Spring | *The English Landed Estate in the Nineteenth Century: its Administration* (1963). |
| F. M. L. Thompson | 'English Great Estates in the Nineteenth Century, 1790–1914' *First International Conference of Economic History: Contributions* (1960), pp. 385–97. |
| F. M. L. Thompson | *English Landed Society in the Nineteenth Century* (1963). |

# 7 High Farming

The years which have been labelled those of 'high farming' stretched from near the beginning of Victoria's reign to the onset of the great depression. It was a period heralded by unmistakable portents of progress—the foundation of the Royal Agricultural Society of England in 1838 and the Cirencester Agricultural College in 1845; the publication of Justus von Liebig's *Organic Chemistry in its applications to Agriculture and Physiology* in 1840 and the establishment of the Rothamsted agricultural research station by Sir John Lawes in 1843; the appearance of a host of new and improved implements and machines, ploughs, harrows, cultivators, tedders and reapers, and not least of Reade's tile drain pipe and of Scragg's and other pipe-making machines in the early 1840s followed shortly by Fowler's mole plough for draining—all these were the signs of the times. Progressive farmers could now learn of the best approved practices in the Royal Agricultural Society's *Journal* and could send their sons to Cirencester for a practical training (although few did); they could visit shows—20,000 visitors attended the first of the Royal Agricultural Society's shows—inspect the prize stock and wonder at the latest advances in machinery; their land could now be cheaply drained; and on their drier and more easily worked soil they could sow the improved strains of seeds and profitably lay (with a manure drill) guano from Peru, nitrate from Chile, and from the 60's, potash from Germany, as well as home-produced superphosphate and basic slag. A new age had appeared, of carefully-controlled breeding, calculated feeding and scientific soil treatment, supplemented by a wide range of mechanical devices and steam power, and the solution to age-old problems of farming were now almost within grasp.[1]

[1] See Philip Pusey's 'On the Progress of Agricultural Knowledge during the last eight years' *J.R.A.S.E.* XI (1850) pp. 381–442.

'High farming' was synonymous with high production, achieved by the judicious application of the new knowledge and equipment available to farmers. The value of higher output was now made abundantly clear, as Caird pointed out, by the growth of urban markets which could be tapped with the speed and economy of the iron road. By the 1850s the railways were already penetrating Scotland and the more distant English counties like Norfolk and Lincolnshire, and offering farmers in formerly remote areas easy access to the teeming populations of London, and the industrial Midlands and north. Coastal steamers and the railways between them made it possible to move farm produce hundreds of miles as cheaply as carrying it 20 or 30 miles by road to the nearest market hitherto; the droving trade dramatically declined as railways penetrated first the fattening and subsequently the more remote rearing regions, and beasts were brought to market with less loss of weight and in better condition; the farmer's transport expenses for his fertilizers, feed, lean stock, seed and implements were greatly reduced; and especially after 1866, when outbreaks of cattle plague had decimated the old insanitary 'town dairies' of London, fresh milk was increasingly brought into the capital by the iron rail.[1] So rapid was the expansion of the market for liquid milk that the production of butter and cheese was adversely affected, making more room in the British market for the long-expanding imports of French and Dutch butter and cheese from America. Imports of foreign dairy produce had been stimulated when duties on their entry had been reduced in 1846, at the time of the Corn Law Repeal (along with the abolition of duties on live animals, meat and hams); and were further encouraged when the remaining duties were abolished in 1860. Not only did imports of dairy produce and meat increase in the third quarter of the century, but so also did imports of feeding stuffs such as maize and cotton-seed cake, which dairymen needed to supplement their own grass and hay.

With rising living standards the urban demand for meat and dairy produce was growing fast, and said Caird, indicated the most profitable direction for farming to follow. Even the arable farmers of the east and south could take comfort from the growing demand for drink-corn, and look to high farming as a substitute for protection. The farmer's business, he claimed, 'is to grow the heaviest crops of the most remunerative kind his soil can be made to carry, and, within certain limits of climate which experience has now defined, the better he farms the more capable

---

[1] For a valuable discussion of the growth of the milk trade in London, see E. H. Whetham, 'The London Milk Trade, 1860–1900' *Econ. Hist. Rev.* 2nd ser. XVII (1964–5).

his land becomes of growing the higher qualities of grain, of supporting the most valuable breeds of stock, and of being readily adapted to the growth of any kind of agricultural produce, which railway facilities or increasing population may render most remunerative. In this country the agricultural improver cannot stand still. If he tries to do so, he will soon fall into the list of obsolete men, being passed by eager competitors, willing to seize the current of events and turn them to their advantage.'[1]

But it was not every farmer who was capable of acting on Caird's advice. Many were still too small, too poor and ignorant, too wedded to the ways of their ancestors. In the weald of Sussex, it was reported, small farms were occupied by men of insufficient capital, men who clung to their two crops and a fallow for fear that any improvement would bring increased rents. In Gloucestershire teams of oxen, driven by labourers of legendary lethargy, took a seven-hour day to plough three-quarters of an acre, or even less; indeed, the progress was so stately, as an observer said, that 'many times have I been compelled to look at some tree at a distance to ascertain whether or not the plough-teams were moving.' And while in that county many signs of improvement were becoming evident in 1850— one-horse carts replacing wagons, hedges being cut down, grain drilled, and ploughed with two or three horses instead of four of five, wider use made of roots and of stall and yard feeding of livestock, drains laid, bones applied as manure, and new farm buildings springing up—yet all these developments needed much extension to become general practice.[2]

Caird himself pointed out numerous instances of highly progressive and grossly antiquated farming continuing virtually side by side. He could go straight from Philip Pusey's advanced experimental farm in Berkshire to those Surrey farmers who employed implements 'of the rudest kind' and who were 'scarcely able to sign their own names'. Cobbett before him had expostulated on the stupidity of the Surrey men who opposed row-culture for turnips on the counts of greater expense and loss of ground. 'How can there be ground lost if the crop be larger?' asked Cobbett, 'and as to the expense, take one year with another, the broadcast method must be twice as expensive as the other.' On the Sussex Downs near Brighton Caird saw wooden ploughs drawn by teams of six oxen, 'an operation which could have been infinitely more cheaply executed by one man and two good horses.' The small tenant farmers, he found, often lacked sufficient capital and were too ignorant to farm properly, and even the Cirencester college was not enlightening them, for its students all came from the

[1] J. Caird, *English Agriculture in 1850–51* (1852) p. 503.
[2] *J.R.A.S.E.* XI (1850) pp. 81–3, 149, 163–4, 168–76.

landowning and professional middle classes and not from the ranks of the farmers.[1]

Caird was prepared to lay part of the blame on the landlords for their fixing of inappropriate rents and inadequate investment, their appointment of unsuitable agents, and the lack of leases, but it is clear that the structure of farming, with its predominance of medium-size and small farms, was an important and perhaps more significant factor. Although the numbers of small owner-occupiers were diminishing, and their land absorbed into larger tenanted farms (as in Westmorland, where Lord Bective bought 25,000 acres from 220 different owners, and in Kirkby Lonsdale 218 small owners were reduced almost by half in 50 years),[2] yet many small farms and holdings survived in every part of the country. According to the figures of 1851 a fifth of the cultivated acreage was taken up by small men with under 100 acres, and such cultivators had a clear majority (62 per cent) of all occupiers of 5 and more acres of land. The farms really well suited for high farming, those of more than 300 acres, numbered less than 17,000 and occupied just about a third of the acreage. The 'average farm', which of course means very little when holdings stretched from 5 acres to over 1,000 acres, was 111 acres.[3] By the 1880s there had been some consolidation, but not very much, for the cultivators of from 5 to 100 acres still accounted for over 60 per cent of all occupiers of 5 and more acres of land. Even Norfolk, the recognized home of the large arable farm, had more farms of between 20 and 100 acres than of over 100 acres, and in other counties of numerous large arable farms, Suffolk, Essex, Cambridge, Huntingdon and Lincoln, the numbers were about the same in each category. In the west and the midland counties largely given over to pasture, such as Leicestershire, Northamptonshire and Warwickshire, there were generally few farms of over 300 acres, and again a rough equality between farms of below and farms above 100 acres. England was still in the main a country of small and medium-size farms, and this unchanging fact clearly limited the application of the methods of high farming.

The connection between advanced farming and the ability, energy and knowledge of the farmer was particularly noticed by a Frenchman, Hippolyte Taine, when he visited England in the 1860s.

> . . . we stopped at a model farm. No central farmyard: the farm is a collection of fifteen or twenty low buildings, in brick, economically designed and built.

[1] *Ibid.* pp. 37, 107–12, 123–4, 127–8; Cobbett, *op. cit.*

[2] F. W. Garnett, *Westmorland Agriculture 1800–1900* (1912) p. 251.

[3] Clapham, *op. cit.* I p. 451; II pp. 263–4.

Since the object was to put up a model, it would not have done to set the example of a costly edifice. Bullocks, pigs, sheep, each in a well-aired, well-cleaned stall. We were shown a system of byres in which the floor is a grating; beasts being fattened remain there for six weeks without moving. Pedigree stock, all very valuable. One bull and his family were Indian and reminded me of Buddhist sculptures. Steam-engines for all the work of the arable land. A narrow-gauge railway to carry their food to the animals; they eat chopped turnips, crushed beans, and 'oil cakes'. Farming in these terms is a complicated industry based on theory and experiment, constantly being perfected, and equipped with cleverly designed tools. But I am not a competent judge of such matters, and amused myself by watching the farmer's face: he had red hair, a clear complexion but marbled with scarlet like a vine leaf baked by the autumn sun; the expression was cold and thoughtful. He stood in the middle of a yard in a black hat and black frock-coat, issuing orders in a flat tone of voice and few words, without a single gesture or change of expression. The most remarkable thing is, the place *makes money*, and the nobleman who started it in the public interest now finds it profitable. I thought I could see, in the farmer's attitude, in his obviously positive, attentive, well-balanced and readily concentrated mind, the explanation of this miracle.[1]

A large proportion of the heavy population of small farmers cultivated unyielding soils with inadequate resources, and moreover, they were often men of little education, suspicious, prejudiced and resistant to new ideas. It was not surprising, therefore, that the progress of high farming was in the eyes of its advocates erratic and unsatisfactory. And of course it was relevant, too, that the properties and use of the new 'artificials' were but imperfectly understood, that the early farm machines exhibited numerous defects, being often badly designed, too heavy and easily damaged, and that much of the pipe-laying was badly executed, the new farm buildings badly sited. Yet where favourable conditions existed of progressive landlords, enlightened tenants, large farms and good markets, the achievements were often remarkable. A growth of mixed farming, with emphasis on fattening, developed on heavy lands, and from the early 1840s yields of wheat were consistently much higher than previously, reflecting the effects of increased application of artificial fertilizers, dung from cattle fed on purchased feeding stuffs, and improved drainage.[2]

Imports of guano rose from under 2,000 tons to 300,000 tons in six

[1] *Taine's Notes on England* (trans. E. Hyams, 1957), p. 132.
[2] M. J. R. Healy and E. L. Jones, 'Wheat Yields in England, 1815–59' *Journal of the Royal Statistical Society* vol. 125 (1962) pp. 577–8.

years after 1841, and the value of bones imported for making super-phosphate rose from under £15,000 in 1823 to over £250,000 within 14 years. More impressive still was the heavy expenditure on drainage, following the cheapening of pipe production and the government's offer of special loans, a sweetening of the bitter pill of 1846; and the expenditure on buildings also rose, encouraged by the removal of the tax on bricks in 1850.

Drainage was the great improvement of the age. If properly executed, it enabled the farmers of heavy and ill-drained soils to cut their costs of cultivation, speed up their operations, and follow the trend towards mixed farming with the introduction of root breaks into their rotations; it enabled machinery to be used, and better advantage taken of the new fertilizers (whose effectiveness and quality were being improved by the invaluable scientific work of Lawes and Gilbert);[1] and the published accounts of the celebrated pioneer farmer, J. J. Mechi of Tiptree in Essex, showed how drainage and intensive farming of poor clay soils could increase profits.[2] In 1851 the Royal Agricultural Society's *Journal* featured a report on the machinery and implements displayed at the Great Exhibition and gave special emphasis to the improvements in tile-making machines and Fowler's draining plough:

Twelve years ago draining-tiles were made by hand, cumbrous arches with flat soles, costing respectively 50s. and 25s. per 1,000. Pipes have been substituted for these, made by machinery, which squeezes out clay from a box through circular holes, exactly as macaroni is made at Naples, and the cost of these pipes averages from 20s. down to 12s. per 1,000. The old price was almost prohibitory of permanent drainage, excepting where stones were at hand: the new invention has reduced this permanent improvement to a rate of £4 or £3 per acre, not exceeding in cost the manure given to a single turnip crop in some high-farmed districts. The result has been obtained by a most spirited competition among mechanists, as no less than 34 different tile-machines competed in 1848 at the York meeting. . . . But for the American Reapers, Mr Fowler's draining plough would have formed the most remarkable feature in the agricultural department of the Exhibition. Wonderful as it is to see the standing wheat shorn levelly low by a pair of horses walking along its edge, it is hardly, if at all, less wonderful, nor did it excite less interest or surprise among the crowd of spectators when the trial was made at this place, to see two horses at work by the side of a field,

[1] See Orwin and Whetham, *op. cit.* pp. 29–31.
[2] *Ibid.* pp. 126–7.

on a capstan which, by an invisible wire-rope, draws towards itself a low framework, leaving but the trace of a narrow slit on the surface. If you pass, however, to the other side of the field, which the framework has quitted, you perceive that it has been dragging after it a string of pipes, which, still following the plough's snout, that burrows all the while four feet below ground, twists itself like a gigantic red worm into the earth, so that in a few minutes, when the framework has reached the capstan, the string is withdrawn from the necklace, and you are assured that a drain has been invisibly formed under your feet.[1]

But drainage schemes were held back by the small size of many farms, the confused pattern of land occupation and ownership which cut across natural strata and water levels and outfalls, and not least by the limited resources of many of the heavily encumbered estate owners. Even after the invention of the cheap machine-made cylindrical pipe in the 1840s, drainage remained an expensive business, which at a minimum expenditure of £4 per acre was much more costly than most of the enclosure schemes carried through 40 or 50 years earlier. It was the government and its cheap drainage loans of 1846 and after—a partial compensation for the loss of protection—which really brought drainage to the enthusiastic notice of land-owners, their agents, and their tenants. The government loans were offered at the moderate interest rate of $3\frac{1}{2}$ per cent and were repayable over a period of 22 years, and the schemes made possible by these loans were to be approved and carried out under the supervision of the permanent Inclosure Commission appointed under an Act of 1845.

The loans were a great success, and were soon supplemented by much larger sums made available by private land drainage companies, who still made use of the expert advice and knowledge of the Inclosure Commission. By 1878 some £12,000,000 had been borrowed from the Treasury and the private companies for drainage schemes in the whole of Britain, and in addition many landowners had found large sums from their own resources. Drainage meant the expansion of mixed farming and heavier stocking of land, so that drainage schemes often involved large additional expenditure on elaborate new dairy parlours, cattle houses, pig-sties and barns. The increased expenditure on buildings reflected the growth of stall feeding of livestock too valuable to be left exposed to the elements, and also a greater appreciation of the value of conserving animal manure. By 1882 over £3,500,000 had been spent on farm buildings in schemes approved under the various Drainage and Improvement Acts, and over

[1] *J.R.A.S.E.* XII (1851) pp. 638-9.

£850,000 on cottages. Clapham estimated that altogether over the 30 years after 1846 landowners poured some £24,000,000 into drainage and other improvements on 4,000,000 or 5,000,000 acres of land.[1]

Much of the new investment never paid. The interest charge and repayment of capital amounted to an annual outlay of $6\frac{3}{4}$ per cent on the sum borrowed, and many landowners must have started too late to clear off the debts before the fall in rents after 1879. Even in the 'golden age' it proved impossible on many large estates to raise rents sufficiently to recoup the expenditure. Nearly £1,000,000 invested in the Duke of Northumberland's estate, for instance, yielded in increased rent only as little as a $2\frac{1}{2}$ per cent return.[2] The same could probably be said about some farmers' investment in the more expensive types of machinery. However, only a few of the large farmers were sufficiently enthusiastic to involve themselves in the expense and pioneering problems of steam cultivation, and in 1867 only 200,000 acres—a negligible proportion—was steam-tilled. The main burden of the great enthusiasm for expensive and largely unremunerative improvements fell on the landowners.

High farming meant intensive farming, producing the highest output per acre through the application of the recent techniques of drainage, fertilizing, feeding, rotating and mechanizing. Its advocates strove after high technical efficiency, and often failed to appreciate that this was not necessarily the same thing as economic efficiency. The weakness of high production based on heavy capital outlays was that it depended for success on growing markets and a high level of prices. Now, it happened that, except for the serious but short depression of 1848–53, the 30 years after the Repeal of the Corn Laws offered such conditions. Markets grew rapidly with the growth of population and of towns now more easily and cheaply reached by railways, and prices for most agricultural produce rose gradually by from 20 to over 50 per cent. Only the price of wheat failed to rise in these years.

After 1853, therefore, market conditions were generally favourable for high farming, for although agriculture was now subject to free competition from abroad in grain, meat and dairy produce, only wheat entered the country in such large quantities as completely to offset the growth of home demand and so adversely affect prices (see fig. 5). Between the 50's and the early 70's the proportion of imported wheat in the total consumption

[1] Orwin and Whetham, *op. cit.* pp. 195–6; Clapham, *op. cit.* II p. 271; Spring, *op. cit.* p. 176. On the development of the legislation designed to encourage drainage and other improvements, see Spring, *op. cit.* Ch. V.
[2] Thompson, *English Landed Society* p. 250.

rose from a little over a quarter to nearly a half, while the retained wool imports more than doubled, and by 1880 greatly exceeded the home supply. In wheat and wool, therefore, English farmers saw a major part of their market gradually fall into the hands of overseas farmers. This did not check the general prosperity, however, for the price of wheat did not fall very far below the pre-1846 level, and wool prices were highly profitable for much of the period, while the overseas exporters of grains other than wheat, and of livestock and dairy produce, although increasingly expanding their consignments, still obtained only a small share of the

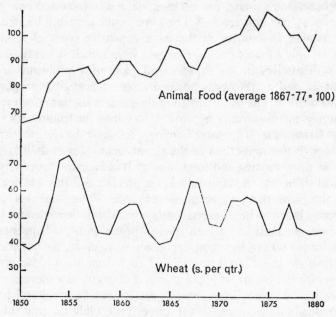

5  *Wheat and Animal Food Price Movements 1850–1880*

market. Under these conditions the English farmer could normally do well, as the rising rentals showed.

The 'golden age' was not uniformly prosperous, however, and the not infrequent years of extreme weather conditions produced severe difficulties for many farmers. The sharp depression following the Repeal of the Corn Laws was unpleasantly complicated by inclement seasons. In 1849 wheat dropped to 44s. 3d., in 1850 to 40s. 3d., and in 1851 to 38s. 6d., the lowest average price for 70 years. Barley and oats also fell, although less severely, and livestock prices fell quite sharply. The dry summer of 1850 and the consequent lack of feed forced large numbers of cattle and store sheep on to the market, while the autumn of 1852 and much of 1853 were so very wet as greatly to reduce the sown acreage of grain and give rise to severe outbreaks of liver rot among the sheep. These years provided a poor introduction to the new era of farming under free trade, but some other years of the period were as bad or worse. 1859–60 saw a very severe winter and late spring with the usual shortage of fodder and glut of lean stock which followed such conditions. Summer rain spoiled the harvest in 1860 and caused outbreaks of sheep-rot, and a shortage of fatstock forced up meat prices. In 1864, by contrast, it was exceptional drought that reduced the supply of feed, and similar conditions were experienced in 1868 and to a lesser extent in 1874.[1] In 1865–6 and 1877 outbreaks of cattle plague (rinderpest) and pleuro-pneumonia appeared. These were so severe in their effects that the government resorted to restrictions on the movement of cattle and paid compensation to the owners of those beasts slaughtered to check the spread of infection, thus reviving methods of dealing with cattle disease first tried during the cattle plague of the middle eighteenth century.

The importance of weather conditions for the harvest and the supply of feed, and the frequent outbreaks of animal disease, point to two of the most serious weaknesses of high farming—the continued, if somewhat reduced, dependence of farmers on the weather, and the failure to advance veterinary science (and the treatment of plant diseases) in line with the general progress of agriculture. Imports now supplemented feed supplies, and drainage reduced considerably the liability of heavy soil farmers to losses of crops and stock, but even by 1880 much land still remained to be drained, and some other land had been inefficiently drained, the pipes too small, or laid at the wrong depth, or subject to some other defect. Not all farmers took advantage of machinery to speed up their operations and enable them to make the best of good weather, and even in 1880 some farmers still got in their harvest by the sickle as in medieval times. Lastly, it was only slowly that some landlords responded to the need for more investment and improved management of their estates, and it was not until 1875 that the tenant's customary right to compensation for his permanent improvements when he left a farm was given legal recognition.[2]

The bad harvests of the middle and later 1870s produced the first clear sign that the degree of natural protection of British farming provided by

[1] Jones, *Seasons and Prices* pp. 103, 169–72.
[2] See Orwin and Whetham, *op. cit.* pp. 153–5, 170–3.

distance was passing away, although contemporaries were misled by the unparalleled harshness of the weather and the unusually high incidence of animal disease. The ominous development was that, despite the shortfall in home production, wheat prices now failed to rise to compensate farmers for their bad crops. For some time the year-to-year oscillations in wheat prices had diminished, as imports took an increasing share of the market. Now the imports had become so large and had such an influence on prices as to make English arable farming disastrously unprofitable when the crop failed. The average price of wheat in the five bad harvests of the 1870s was under 50s., while in the five worst years of the 1860s it had been about 60s.—and in 1879, the worst harvest of the century, wheat sold for only 43s. 10d. Worse was yet to come, for in the 80's wheat prices continued to drop and only reached their lowest point in the 90's. Furthermore, imports of butter, cheese, eggs and bacon began to rise sharply in the late 60's, and although prices were not affected for some years, pasture farmers, too, began to feel the pressure of competition. The 'great depression' revealed that the conditions of the 50's and 60's had been fortuitous. Free trade had not ruined the English farmer because his potential competitors overseas were not yet fully in a position to bring their natural advantages and lower costs of production to bear. For some 30 years after 1846 the English farmer was protected by an adventitious set of circumstances—by wars, transport costs, and technical problems which prevented American and Russian farmers from fully exploiting their fertile soils.

By the 70's, however, the American Civil War was past and the railway was penetrating the great plains; and Russia, an old exporter of grain to western Europe, was presently going through a period of low shipments but was soon to become important again. The American homesteader could now obtain barbed wire—the cheap solution to the vital problem of keeping cattle out of his wide prairie acreage—and could use increasingly efficient machines to sow, reap and thresh his crops, thus releasing him from the limitations of scarce manpower; the 'elevator system' and trading in 'futures' made for cheap, rapid and profitable handling of grain in bulk; and the railways and the revolutionized steamships carried the cheap produce across half a continent and the Atlantic. Between the 1870s and the end of the century the cost of carrying a quarter of grain from Chicago to Liverpool fell by nearly three-quarters, from 11s. to under 3s. By the 70s's also, the English dairymen were feeling increased competition, not only from American cheese, but also from the produce of their European neighbours, the Dutch and the French, and from the

Danes, who were busily engaged in turning their rather unpromising countryside into a huge, well-organized factory for the production of cheap but high-quality butter, cheese, lard and bacon. Only milk, hay and straw, which represented something over a fifth of the value of English agricultural production in the 1870s, completely escaped foreign competition.

The fall in prices in the last quarter of the nineteenth century was not universally disastrous. It hit earlier and more severely the arable farmers, and of those, the occupiers of the heavy clays, often still undrained, with their lower yields and high costs, came off the worst. The pasture farmers and their landlords were much less affected because the market for their produce was growing much faster than was that for grain, and because the foreign competition in meat, cheese, butter and bacon, although on an entirely new scale, still did not have a very severe effect. While the prices of imported meat and dairy produce fell by 20 per cent or more, the expanding market and shift on consumers' tastes meant that the prices of the home produce fell by only half as much. But taken as a whole, the great depression in agriculture, with its bankrupted tenants, unemployed labourers and straitened landlords, was real enough. The future for English agriculture lay in reducing costs and becoming more competitive, and in converting a large proportion of arable land to pasture for dairying and beef production. The heavy investment in drainage and buildings came to an end, and inessential maintenance was skimped. Low costs became more important than high output, and the age of high farming was over. Yields on arable land continued to rise, however, partly because of the conversion to pasture of the poorer arable soils, and partly because of greater availability of artificials and town manure. But in general, the change of conditions from the 1870s showed that after all Caird was wrong: high farming was not a substitute for protection when English agriculture was exposed to the full force of a competition undreamed of in 1846.

The failure of English farmers to perceive at first the full effects of cultivation of virgin soils and the transport revolution, and their sluggishness in adjusting themselves to the new market situation, have given rise to criticisms of inflexibility of outlook, rigidity of production methods, and misdirection of investment. English landlords and farmers, it has been complained, clung blindly to corn, and even invested large sums in expanding arable output at the very time when an unprotected market threatened to make intensive methods of arable farming quite uneconomic.

A re-examination of the period of high farming, however, has shown that these criticisms need to be greatly modified. Apart from the severe outbreaks of sheep-rot, rinderpest and pleuro-pneumonia, and the years of exceptional weather conditions which have been mentioned, between the early 1850s and the middle 1870s agriculture was generally and increasingly profitable. It has been estimated that the farmers' 'incentive income', their reward for undertaking the risks and management of farms, almost doubled in this period.[1] The major part of the improvement in profits, however, was being earned by the producers of meat, dairy produce and vegetables. Following his tour of 1850–1, Caird pointed to the advantages of pasture by calculating that while the price of wheat was about the same as it was in Arthur Young's time in 1770, the prices of butter, meat and wool had nearly doubled. Rents reflected this difference, he said, for they averaged 31s. 5d. per acre in 15 midland and western counties as against only 23s. 8d. in 18 eastern and southern counties. His map (see fig. 3, page 142) showed the division between the profitable grazing counties and the less profitable corn counties.

Caird also drew attention to the very significant point that this difference was likely to become even greater in the future. With improving standards of living the pattern of demand was shifting away from bread and towards a higher consumption of meat and dairy products. The farmer himself spent more on meat than on bread, said Caird, and this change had been gradually spreading. 'With the great mass of consumers, bread still forms the chief article of consumption. But in the manufacturing districts where wages are good, the use of butcher's meat and cheese is enormously on the increase; and even in the agricultural districts the labourer does now occasionally indulge himself in a meat dinner, or seasons his dry bread with a morsel of cheese . . . the great mass of the consumers, as their circumstances improve, will follow the same rule. . . . Every intelligent farmer ought to keep this steadily in view. Let him produce as much as he can of the articles which show a gradual tendency to increase in value. . . . With the present prices, and the knowledge of the fact that the rich corn provinces of the continent are open to us, and are daily becoming more accessible by the extension of railways and steam navigation, there seems good reason to anticipate the permanence of a low range of prices. The safe course for the English agriculturist is to endeavour, by increasing his livestock, to render himself less dependent on corn, while he at the same

[1] J. R. Bellerby, 'National and Agricultural Income 1851' *Economic Journal* LXIX (1959) p. 103.

time enriches his farm by their manure, and is enabled to grow heavier crops at less comparative cost.'[1]

Although some farmers were too small, too lacking in capital, or too conservative and short-sighted to depart from the ways of their forefathers, many others followed Caird's advice. The cereal acreage was at its peak about 1855 and then gradually began to decline. By 1870 the total arable acreage in England and Wales (including 2,000,000 acres of root crops and over 3,000,000 acres of rotation grass) was 14,849,000 acres, while the area of permanent grass amounted to 11,108,000 acres. Wheat was still by far the largest single crop with over 3,250,000 acres; but even so, cereals together occupied only a little over half the total arable acreage, and less than one-third of the total farmland.

The total acreage described as 'arable' is thus misleading, since well over a third of it was taken up by root crops, green crops and rotation grass used for the support of the flocks and herds in the various systems of mixed farming. Even within these systems, moreover, there was a growing tendency towards greater concentration on livestock production. While much of the rather moderate increase in permanent pasture after 1850 was taken up for dairying, more important for meat production was 'the growing emphasis on the livestock enterprises within the mixed farming systems which were so popular, and the acceptance by farmers that their grain crops might more profitably be fed to fattening stock than sold for human consumption'.[2] Although permanent pasture was on the increase, there was insufficient grassland of good fattening quality to meet the rising demand for meat, and temporary grasslands or leys were not thought to be as productive of fat cattle as arable farms. The arable farmers, therefore, became important producers of fatstock. 'In the 1850s and 60's stall-feeding and yard-feeding of cattle was intensified in mixed farming regions, and spread into districts such as the dairying parts of Cheshire and Gloucestershire, where it was hitherto unknown, on to chalk and limestone uplands where until the 1840s sheep had been almost the only stock, and into parts of Cornwall and Cumberland where the entire dependence had previously been on grain.'[3]

'High farming' could be used in Caird's sense to mean a high outlay of

---

[1] Caird, *op. cit.* pp. 476, 480-9.
[2] E. L. Jones, 'English Farming Before and During the Nineteenth Century' *Econ. Hist. Rev.* 2nd ser. XV (1962-3) p. 149.
[3] E. L. Jones, 'The Changing Basis of English Agricultural Prosperity 1853-73' *Ag. Hist. Rev.* X (1962) pp. 112-13.

capital to achieve a high output per acre—by means of artificials and stocking with good quality breeds of cattle, drilling of the best seeds and the use of machinery to harvest and thresh the crops, together with his requirements of thorough drainage, removal of 'unnecessary obstructions to economical tillage', 'convenient farm-roads', and 'well-arranged buildings'. This, we have seen, was being done on the progressive land-lords' estates in the middle decades of the century, and indeed the change reflected the influence on English farming of both the progressive large-scale East Lothian husbandry and the Scottish agents now widely employed in managing English estates—an interesting reversal of the eighteenth-century pattern of agricultural influence which then, of course, ran from south to north. But 'high farming' in a narrower technical sense, meant 'high feeding', a more intensive development of the Norfolk four-course mixed farming system, with its succession of fodder and grain crops, its arable flock, and yard-fed bullocks. High feeding meant an increased output of all the products of the system, as the livestock, lavishly fed on imported oilcake, produced more meat and manure, and the ploughland, now enriched by heavier dressings of manure and arti-ficials—especially important for the light soils where the Norfolk system predominated—produced higher yields both of grain for the market and of fodder crop for the stock. When the Corn Laws had gone, wheat was quite often cheap enough to be fed to bullocks and pigs on a large scale. Indeed, this had occasionally been the case even before 1846, and some farmers believed there was greater profit in feeding wheat to stock than selling it on a market depressed by imported corn. In this way mixed farming gained a greater flexibility in face of the lower prices for wheat, and when corn was low farmers could hope to gain from larger sales of fatstock.[1]

In the middle decades of the nineteenth century, therefore, a subtle change was occurring in the character of much of the mixed farming, and as the margin between grain and livestock prices widened so the emphasis tended to shift to meat rather than corn as the end product. This was accompanied by a greater variety in the types of roots and grasses sown, in order to avoid the blight which often affected the turnips and the inci-dence of 'clover-sick' soil. There was also an increasing irregularity in the succession of crops. The best farmers did not cling inflexibly to the Nor-folk four-course but experimented and changed about endlessly. The large farms of the Scottish Lothians, the Carse of Gowrie and southern

[1] E. L. Jones, 'The Changing Basis of English Agricultural Prosperity 1853–73' *Ag. Hist. Rev.* X (1962) pp. 104–8.

Perthshire, were now the recognized home of the best mixed farming; this was not only because of thorough drainage, 21-year leases and the use of machinery, but also because of the flexibility of the rotations. A striking feature of Scottish farming was the growth of sheep-raising and fattening on turnips. This practice spread in the later eighteenth century with the introduction of the southern breed of black-faced Linton sheep, and was particularly influenced by the high prices of the Napoleonic Wars. High prices and good profits gave the farmers the capital necessary for their new rotations and heavier stocking. Quite soon, the sheep spread into the old areas of cattle raising, Dumfriesshire, Perthshire, southern Argyle and Morven. The Scottish cattle trade reached its peak about 1835 and then declined in favour of the traffic in sheep. By 1850 the sheep trade had surpassed that in cattle at the Falkirk tryst, and continued to grow until about 1870 when it was in turn affected by falling prices for wool and mutton, and deterioration of the highland pastures.[1]

In England the desire of enterprising farmers to experiment with new crops and fertilizers, to grub up hedges for machinery, and to vary their rotations in accordance with market trends, was an important reason for the decline of long leases in some areas. In Wiltshire, for example, many farmers gave up leases unilaterally, and were even opposed to the idea of introducing compulsory compensation for improvements or 'tenant right', while where leases remained they were often inoperative and seemed 'only to be binding on landlords'.[2]

It is clear then that, even before the collapse of wheat prices after 1874, English farming was already moving away from its traditional reliance on wheat as the prime market crop. The shift of emphasis within mixed farming made it possible for the farmers to take advantage of rising meat prices without greatly reducing their arable acreage, and this was a natural reaction to the market situation which developed in the early years of free trade. Of course, when the full effects of foreign competition were felt, 'high feeding' on arable farms gave way before the unremunerative prices for wheat and the cheapness of imported feeding stuffs, while the more conservative arable farming, the rigid persistence with arable on heavy clays and with the four-course rotation elsewhere, went on to disaster and bankruptcy. It might even be argued that when the great slide in grain prices began, the capacity of mixed farming for adaptation in a

[1] Haldane, *op. cit.* pp. 192–4, 200–6.
[2] R. Molland, 'Agriculture, *c.*1793–*c.*1870' *V.C.H. Wiltshire* IV (ed. E. Critall, 1959) p. 75.

period of rising livestock prices intensified the subsequent depression in the eastern half of the country, for many farmers tried to carry on as before, confident in the ability of so flexible and productive a farming system to survive.

But there were, indeed, a number of other reasons why the retreat of arable farming was so stubborn and prolonged. In the first place, farmers were impressed not only by its adaptability but also by its greater total productiveness, as compared with livestock-fattening on pasture. Furthermore, their landlords' recent heavy expenditure on drainage and buildings, and the farmers' own outlays on improvements, encouraged a confidence in the permanence of the system. There were undoubtedly some farmers, too, who felt with Cobbett that the only farming that deserved the name was that which ploughed, dragged, harrowed and rolled the soil, and by the sweat of the brow produced its 'five to six quarters of wheat standing upon the acre, and from nine to ten quarters of oats standing along side of it'.[1] And in the early years of the depression it was difficult, of course, to be sure that the bad times were not merely a temporary phase, due largely to bad weather and animal distempers, and were really the consequence of a fundamental and permanent shift in agricultural conditions. In the early 1880s, for instance, some agricultural writers were telling their readers that this was a very good time to think of buying a farm.

Lastly, the long unprosperous years of waiting to see if things would improve—an optimism encouraged by the two slight upward movements of wheat prices in the early 80's and early 90's—robbed the farmers of initiative and capital, and both were important for making the transformation to pasture. For the most part adjustment required the new capital and enterprise of a new generation of farmers, so that in 1896 it was remarked by an observer in Essex that 'newcomers are going in for milk, cheese, butter, fruit and sheep, but with the average Essex farmer it is corn, corn, corn'.[2] The rooted faith in the old rotations and the technical achievements of high farming was not something that the pre-depression race of farmers could lightly cast aside.

Investigation of the migration from the rural areas of England and Wales to the towns and overseas has shown that there was a considerable increase in the numbers moving in the 1850s, followed by a check in the 1860s, and then an even higher outflow in the 70's and 80's. Not all of the

[1] Cobbett, *op. cit.* II p. 285.

[2] Quoted by T. W. Fletcher, 'The Great Depression of English Agriculture 1873–96' *Econ. Hist. Rev.* 2nd ser. XIII (1960–1) p. 431.

people who deserted the countryside were agricultural labourers of course, and many migrants drifted back again, repelled by the high prices and the difficulty of finding houses in the towns. But a large proportion never came back, and between 1851 and 1871 the number of agricultural workers fell by over 250,000, or 22 per cent. Such a large loss of labour could not but be felt, even in the over-populated counties of the south and east which, in fact, supplied most of the migrants at this time. The reduction in the labour force meant a greater shortage at haytime and harvest, an additional inducement for farmers to use tedders, reapers, threshers and other machines, and greater resort to the labour of gangs of women and children. In addition, farmers did more to try to keep their men on the land, offering better cottages, more allotments, improved schools and other ameliorations of the labourer's lot.

The pressure on the farmers was increased by the continued expansion of the demand for labour as cultivation still extended into waste areas, while high farming, with its intensive methods of cultivation and of fattening, and its higher yields, was dependent on an adequate labour force. Even where machinery was much employed there was often a consequent need for more men and certainly a need for more horses, and the increase in the numbers of horses, both in country and in town, required more men to rear, drive and feed them. Average money wages rose by nearly 40 per cent between the early 1850s and the early 1870s, but the increase was uneven and the rise in real wages was rather less impressive. The west and south-west was as yet little affected by the movement from the land, but generally the improvement in the economic position of the labourer was reflected in a stronger sense of independence and there were occasional sporadic and small-scale attempts to form unions.[1]

The emigration from the countryside increased the need for more seasonal migration within it. Skilled workers were in demand and often travelled widely as self-employed craftsmen, while the harvesting gangs who roamed the country were not composed solely of Irishmen, Welsh or Scots: there was, for example, a regular transfer every summer of surplus labourers from the dairying areas of north Wiltshire to the south of the county in order to get harvesting work, the men returning to their own villages to spend the winter on drainage or road repairs, on the railways perhaps, or on relief.[2]

The railways had an important influence, not only in cheapening and

[1] See E. L. Jones, 'The Agricultural Labour Market in England, 1793–1872' *Econ. Hist. Rev.* 2nd ser. XVII (1964–5) pp. 327–37.

[2] Molland, *loc. cit.* p. 82.

simplifying travel, but also in providing jobs and raising rural wages, and in fanning the spark of enterprise which among the lowest-paid labourers of the more remote west and south-west burned very low. Education and a widening of horizons were vital when the labourers, like those in north Devon, had no idea of where better employment was available or how it might be reached, and clung to their squalid cottages and 7s. or 8s. a week rather than face the unknown.

Such ignorance was among the difficulties which Canon Girdlestone had to overcome when he undertook to organize the migration of labourers and their families from his parish of Halberton in 1866. 'Almost everything had to be done for them, their luggage addressed, their railway tickets taken, and full and plain directions given to the simple travellers ... written on a piece of paper in a large and legible hand. These were shown to the officials on the several lines of railway, who soon getting to hear of Canon Girdlestone's system of migration, rendered him all the assistance in their power... Many of the peasants of north Devon were so ignorant of the whereabouts of the places to which they were about to be sent [in Kent and the northern counties], that they often asked whether they were going "over the water".'

The Canon's efforts—originating in a letter to *The Times* which attracted much attention and many offers of work and money for the labourers—were spurred on by his discovery of the miserable conditions existing in north Devon. As an addition to the wage of 7s. or 8s. a week, the men were allowed a daily three pints or two quarts of unsaleable and almost undrinkable cider, and the somewhat dubious advantage of purchasing 'grist corn' or low-grade wheat at a fixed price. The normal day was one of $10\frac{1}{2}$ hours, from seven in the morning to half-past five in the afternoon, but it was sometimes longer, and the labourers' wives were paid only 7d. or 8d. a day, and were often compelled to work as a condition of the husband's employment. The diet which labouring families could afford was simple and monotonous enough: a breakfast of 'tea-kettle broth'—a soup made by pouring hot water on some slices of bread, with some seasoning and an onion; bread and cheese for the forenoon and afternoon meals in the field; and a supper of bacon, potatoes and cabbage, with butcher's meat as a very rare treat. The farmers of the district reacted to the Canon's intervention in their affairs by creating opposition in the vestry and absenting themselves from church, but the Canon was not to be so easily deterred. Within six years between 400 and 500 men, many with families, had been moved to high wage areas through this voluntary effort of one individual, and soon the idea spread to the surrounding counties and was taken up by the agricultural labourers' unions in the early 1870s.[1]

It was not entirely by chance, therefore, that migration spread to the south-western counties and reached a peak in the 1870s, while that from Wales began to rise fast. The Welsh migration came to a head in the 80's, with the more stable outflow from the northern counties, and in the 90's the movement from the countryside was generally subsiding, mainly as a consequence of the heavy migration of the previous decades. 'Fewer potential migrants were being born, because the parents who might have reared them had already migrated.'[2] The outflow, although averaging between the 1850s and the 1890s over 75,000 people a year, caused only local and short-term falls in the rural population. The countryside was not depopulated but stood still in numbers, while the towns swelled and the emigrant ships filled.[3] Only the increase in the country population, although often the pick of it, was creamed off by migration, but the movement was sufficient to raise agricultural wages and to encourage farmers to economize in labour by using machinery.

From the 50's large farmers were turning increasingly to a growing range of machines that was now reasonably efficient and was widely advertised. As we saw in an earlier chapter, threshing machines had long before this made an impact on labourers' winter employment, and steam threshing was now very common. Other machines that were becoming widely used, such as horse-drawn drills and cultivators, tended to extend the area of drill culture, and together with some other devices such as the mole plough, probably increased the demand for labour. Drilling and drainage meant better and heavier crops, and to make effective use of machinery fields had to be made larger and more regular. The grubbing up of hedges and levelling of banks and ditches created work, and in areas newly reclaimed from waste, gangs of women and children were employed to remove the stones and weeds and so prepare the land for machine cultivation.

However, there were many travelling threshers, mowers, and drills for artificial fertilizers, which farmers could hire in order to economize in labour. The machines themselves gradually improved, with better designs, more suitable materials and lighter and more robust construction. They also became cheaper, and many of the smaller models could be had

[1] F. G. Heath, *The English Peasantry* (1874) pp. 140–6, 153–6.
[2] A. Cairncross, 'Internal Migration in Victorian England', in Cairncross, *Home and Foreign Investment 1870–1913* (1953) p. 75.
[3] *Ibid.* p. 77.

for as little as £10 or £20. Farmers were increasingly encouraged to buy them as adult labour became dearer, harvesting bands scarcer, and restrictions were imposed on the employment of children by the Gangs Act of 1867 and by the Education Acts after 1870. The scarcity of hands at harvest time encouraged farmers to reduce the risk of rain-damaged crops by hiring or buying the American reapers which had been shown at the Great Exhibition. Mechanical reaping not only saved time but also reduced labour costs (although McCormick's reaper still required a dozen hands to follow the machine and gather, tie and stook the cut corn), and by the 1870s Caird estimated there were 40,000 reapers in use. It was in order to dispense with the heavy expense of plough-teams that some large farmers involved themselves in steam ploughing. One farmer of 1859, we learn, purchased a 14 h.p. steam engine and sold off his seven four-ox teams. He grubbed up hedges and reduced his 36 fields to 9, and claimed to have reduced his costs of cultivation by a third.[1]

By the 1860s the use of machinery had so reduced the summer demand for labour that the wives of labourers were going out to work much less and the children were staying longer at school. At the same time, as we have noticed, wages were rising, and fairly rapidly. The average money wage for the lowest-paid day labourers (exclusive of allowances in kind) was about 9s. 6d. a week in 1850 and about 11s. in 1860; by 1870 it had climbed to 12s., and by 1880 to 13s. 7d., and continued to rise slowly through the depression years. Real wages rose irregularly, rapidly in the 1840s when a fall in food prices increased the labourer's purchasing power perhaps by over 30 per cent, much less rapidly between the 1850s and the 70's when bread and most other provisions were dearer, but rose again substantially in the years after 1874. Money wages, of course, differed greatly north and south of Caird's high wages line (see fig. 3, page 142), and varied, too, from one district to another even in the same region. If anything, it appears from the returns of 1867 that the northern labourer's margin of over a third of the southern labourers' average wage (northern counties 11s. 6d., southern counties 8s. 5d. in 1850), was being reduced in the decades after 1850.

But wage rates have never told the full story of the agricultural labourers' conditions. By working at piece rates on such jobs as hedging, draining, mowing, turnip-hoeing and harvesting, many labourers could earn one or two shillings a week above the day rates, and the skilled men who had care of animals also earned more, although they often worked longer hours for their higher pay. There was also an important group of

[1] Molland, *loc. cit.* pp. 85–6.

independent workers in the countryside whose special skills of hedging, mowing, or thatching, or whose handiness at repairs and odd jobs, and in fetching and carrying, kept them in constant demand. Allowances in kind varied greatly from one area to another, and on average they may have effectively raised the total earnings by perhaps 15 or 20 per cent; they usually included some food and drink, quite often a sack of corn and a patch of potato ground, and sometimes a free cottage and free fuel. The labourer's wife and daughters could earn 6d. or 8d. a day in the fields, or they might work at home, laundering or glove-making, and even the youngest boys might earn a few coppers for scaring the crows. On the other hand, day wages were greatly reduced by wet weather or illness, and the allowances were not always worth very much. Cottage rents were sometimes high in relation to wages, often bearing a proportion of a quarter or more, and by the time they were 50 many of the men were crippled by rheumatism, the result of dampness indoors and of working all day outside in rain-sodden clothes. When the Warwickshire labourers went on strike in 1872 the special correspondent of *The Daily News* summarized the income and expenditure of a typical family: 'Wages, father 12s.; son 3s.; = 15s. per week. The week's bread and flour, 9s. 4d.; one cwt. of coal, 1s. 1d.; schooling for children, 2d.; rent of allotment (1 chain), 1d.; total, 10s. 8d. Leaves for butcher's meat, tea, sugar, soap, lights, pepper and salt, clothes for seven persons, beer, medicine and pocket money, per week 4s. 4d.'[1]

This family apparently had their cottage as an allowance, and Warwickshire was not by any means a county of the lowest wages, nor did the early 1870s represent the nadir of rural conditions. In fact things had been very slowly on the mend for some 20 years past, for in the age of high farming, when machinery and more valuable stock were raising the standards demanded of at least a minority of labourers, landowners and farmers were beginning to have a greater interest in keeping the labourer on the land and making him more efficient. A good part of landlords' investment at this time, as earlier, went into cottages, more cottage gardens and allotments were provided, and village schools were built. The paternalism of the squire and the visits of his daughters to the needy poor with blankets and food may strike us now as a poor substitute for adequate wages, but neither squire nor farmer could do very much about the fundamental cause of low wages so long as the labourer and his numerous progeny were still so thick on the ground and were immobilized by poverty and ignorance.[2]

The living conditions of the labourer and his family varied enormously

[1] Heath, *op. cit.* p. 2.
[2] See *J. R.A.S.E.* XI (1850) p. 754.

from one district to another. If he were unfortunate his wages might be paid very largely in truck, in sacks of corn and potatoes at rates above the market price, in cider unfit for ordinary sale, and even in the carcases of stock that had fallen victim to disease. His cottage would be wretchedly small and badly built, perhaps of mud walls and a thatched roof, as in Wales and parts of the south-west, with only one bedroom, as nearly half of all cottages had in the 1850s, or even with only one room for the whole family—consisting not uncommonly of three generations, all living, cooking, eating, washing and sleeping in the one confined sordid space. Some cottages still had no privies of any kind attached to them, and some had floors of clay or broken stone, like those in Dorset and Somerset, which 'heaved' or became sodden when it rained. In the exceptionally bad village there would be neither allotments nor gardens for potatoes, and as in Joseph Ashby's Tysoe, it might take years of struggle to obtain them.[1] But generally allotments and gardens were becoming more numerous and more easily obtained, although the rents charged for them were often exorbitant. By 1886, at all events, there were as many as 646,000 allotments and gardens of an eighth of an acre or more, 93,000 potato patches in the fields, and over 9,000 'cow gates' or rights to graze a cow on lane verges and spare ground, so that few labourers who wanted it could have been without access to some land.

If the cottager were lucky he would live in a roomy, newly-built cottage erected by a great landlord; and if his landlord were such a one as the Duke of Bedford he would be lucky indeed, for the Duke refused to allow the farmers to have the letting of his cottages, and he did this deliberately in order to weaken the farmers' hold over the labourers. Good estate cottages might be found wherever large blocks of the great owners' property existed. Built of brick and slate, with a touch of the fashionable gothic styling, and with two or three bedrooms, they put to shame the highly-rented slums which still proliferated in the former 'open' villages of small property-owners. The 1865 modification of the Poor Law, which spread the poor rate over the whole union instead of the single parish, encouraged large owners to build, and by the 1880s housing conditions in the countryside had been much improved by their work.[2]

In the north farm servants still lived in with the farmer, as they continued to do beyond the end of the century. In Westmorland they seem to have been fairly comfortable, and from all accounts enjoyed a plain but

[1] See M. K. Ashby, *Joseph Ashby of Tysoe, 1859–1919: a study of English Village Life* (1961).

[2] Clapham, *op. cit.* II pp. 285–6, 507–8.

adequate diet. About 1870 this consisted of porridge for breakfast, and bread and cheese with beer or milk at 'drinking', taken about ten in the morning; for dinner at noon they had a fruit pudding followed by meat with potatoes and vegetables, and for tea at four o'clock bread and cheese with tea, or perhaps bread and butter with jam; and supper rounded off the day with either porridge or bread and cheese again. The bread, it is interesting to find, was still a 'maslin' mixture of two-thirds wheaten flour and one-third rye. From their annual wages of £10 to £12 for men, and £7 to £8 for women, the Westmorland farm servants were able to save, encouraged by the prospect of one day taking a cottage and small-holding. In 1868 the Penrith branch of the Carlisle Savings Bank had deposits totalling £9,259 from 260 male farm servants, and £7,904 from 240 female servants. Many of the servants, and also some day labourers, were members of local friendly societies.[1]

Even in Westmorland the pernicious gang system was in use for such work as spreading manure, for weeding, thinning turnips, and taking up potatoes. The real home of the gang, however, was in the eastern counties, where they were recruited from the surplus men, women and children of open parishes to work for the large farmers in areas where labour was scarce. Starting about the 1820s, the gangs had something in common with the migratory groups of harvesters, hop-pickers, fruit-pickers and sheep-shearers. The eastern counties gangs, however, were employed largely in the monotonous and back-breaking work of lifting potatoes and weeding, and even in picking up stones and couch-grass roots. They were much in demand in the new arable farms made up of land recently enclosed from the fens and wastes, where there were few or no cottages and labourers had to come from five or six miles away—and to employ a gang was cheaper than building cottages and hiring regular labourers. The gang masters had little difficulty in obtaining women and children from the populous villages, for labouring families incurred debts during the slack periods of winter-time which had to be paid off in summer, and the schools were few and poorly attended. In the late 60's a social outcry against the conditions under which the gangs worked and lived, the brutalizing and degrading character of their work, the illiteracy of the children and the immorality of the young people, led to legislation. The Gangs Act of 1867 established a licensing system for the gang masters and forbade the employment of children under the age of eight. But what really crippled the gangs was the coming of compulsory education and the gradual raising

[1] Garnett, *op. cit.* pp. 95–6.

of the school-leaving age under the series of Education Acts which began in 1870.[1]

The nineteenth-century countryside was certainly not the unspoiled haven of rustic simplicity that romantic writers liked to picture. Although conditions varied greatly from one district to another, as Cobbett had noticed in his *Rural Rides*, poverty and ignorance, long days of hard and rough work, and the sordid living conditions, too often fostered an existence which revolved round the beer-shop and the unceasing struggle to make ends meet and keep out of the workhouse. Harriet Martineau, the Victorian radical writer, showed an appreciation of the real nature of rural life when she wrote to her friend Elizabeth Barrett in 1846:

> I dare say you need not be told how sensual vice abounds in rural districts. Here [at Ambleside in the Lake District] it is flagrant beyond anything I ever could have looked for: and here while every justice of the peace is filled with disgust and every clergyman with (almost) despair at the drunkenness, quarrelling and extreme licentiousness with women—here is dear good old Wordsworth for ever talking of rural innocence and deprecating any intercourse with towns, lest the purity of his neighbours should be corrupted. He little knows what elevation, self denial and refinement occur in towns from the superior cultivation of the people. The virtues of the people here are also of a sort different, we think, from what he supposes. The people are very industrious, thrifty, prudent and so well off as to be liberal in their dealings. They pride themselves on doing their work capitally; and in this point of honour they are exemplary.[2]

The growth of savings banks and friendly societies were evidence of the thrift and industriousness on which Harriet Martineau commented. The village schools, too, although generally inadequate, were reducing illiteracy and opening the minds of the young to new ideas and the progress of the wider world. By the 1860s there were clear signs of a change in the labourers' outlook and spirit. Observers remarked that many more of them were now to be found reading newspapers and taking an interest in political affairs. The young men looked down on the conditions of their parents and left the countryside for the towns. They could not for the most part hope for a well-paid job, but at least they could attain a secure one. Accordingly they became policemen or railwaymen, enlisted as soldiers or perhaps emigrated. Those who stayed behind began to show a much more

[1] See Joan Thirsk, *English Peasant Farming* (1957) pp. 217, 268–70.
[2] R. K. Webb, *Harriet Martineau: a Radical Victorian* (1960) pp. 260–1.

independent attitude towards the farmers, striking at harvest times for better pay, and developing a rudimentary trade unionism.

Unionism in earnest began in Kent in 1866, and by 1871 had spread to Buckinghamshire, Hertfordshire, Herefordshire and Lincolnshire. It was concerned with the improvement of wages, emigration, and the movement of labour to the high wage counties such as Staffordshire, Lancashire and Yorkshire. The movement gathered strength in the following year, 1872, when some Warwickshire labourers invited Joseph Arch, a local man, but a champion hedger and mower and a self-educated Methodist lay-preacher, to address a meeting at Wellesbourne. There, on February 14th, under a chestnut tree and in the rain, Arch addressed a crowd of 1,400, and suggested the motto, 'Defence, but not defiance.'

These years, although not perhaps the most prosperous of the era of high farming, were good ones for the labourers' new-found belligerence. A high level of employment in the towns made migration easy, and so the strikes in the countryside proved effective and pushed up the level of wages by 20 or 30 per cent. By 1873 Arch's Union was a national organization which numbered some 150,000 members (not all of them agricultural labourers) out of a total labour force of 650,000, but with the main strength lying in the low wage counties south of the Trent. The growth of unionism was in fact most marked in corn-growing areas with a high proportion of day labourers to farm servants, and where wages were low, but not so low as to have completely demoralized the men. The improvement of education and a wider outlook in the countryside was certainly a force that fostered the movement, and there was a strong element in it of nonconformity. Many of the leaders, like Arch, were Methodist lay-preachers. Some leaders came from outside the ranks of the labourers, from the village shopkeepers, teachers and journalists. Consequently this first widely-organized union movement had a strongly moral, intellectual and political aspect, with much emphasis on education and temperance, on demands for the franchise and the nationalization of the land. Arch's National Union even had its own newspaper, and financial assistance came in from organized workers in the towns, as well as from individual sympathizers. Some landlords, too, were sympathetic; and a few helped the cause by lowering the rents of farmers who raised wages in response to the Union's demands, and by preventing their tenants from joining the hostile farmers' associations.[1]

The tide soon turned, however. The conditions of agricultural employ-

[1] J. P. D. Dunbabin, 'The "Revolt of the Field": The Agricultural Labourers' Movement in the 1870s' *Past and Present* 26 (Nov. 1963) pp. 69–72, 89.

ment—the wide scattering of the labour force in small units, the close personal ties between the farmers and the men, the farmers' hold over cottages and allotments, and the lowness and irregularity of the wages, together with the unfriendly attitude of many parsons and squires and a shortage of men able to play the exacting part of local leaders—were all very unfavourable to permanent Union organization. Moreover, by 1874 the farmers were replying to the men's initiative with lock-outs and evictions from tied cottages, the employment of non-Union labourers and Irish as blacklegs, and a greater use of labour-saving machinery. Indeed, it was only in the great lock-out of 1874 that many farmers first appreciated the advantages of machinery and greater economy in labour. The farmers attacked the radicalism of the Union, arguing that the labourers in their ignorance were the dupes of men of hidden selfish interests, and generally they were unwilling to see any need for a change in the labourer's social and political status. Some farmers did think that higher wages should be paid and that better pay would result in a more efficient labour force; others held that higher wages could only result in the abandonment of the poorer land, a greater use of machinery, and an unwillingness on the part of farmers to keep elderly labourers on at work out of charity; while yet others believed that since wages were fixed by the supply and demand for labour, the unions could have little permanent effect on the wage level.[1]

The labourers were too poor and their Union too weak to be able to resist lock-outs and blackleg labour on a wide scale. Some unions, such as that of the Lincolnshire labourers, had stood aloof from the national organization, and the movement suffered from disunity and internal dissensions. From 1875, too, the demand for labour was adversely affected by the bad harvests and the fall in grain prices, and so the membership gradually dwindled away. By 1879 almost nothing was left of the great enthusiastic surge of seven years before. Arch recognized that the labourers must try another road to independence and a better life, and he turned to the political enfranchisement which was achieved in 1884 and 1918. The rising prices of the late 80's led to some revival of unionism, but it was short-lived and collapsed again in the bad years of 1893 and 1894.

The failure of unionism coincided with the development of compulsory education and a reduced demand for agricultural labour. By the 80's the outflow of men from the countryside was so large that, together with the big fall in food prices, the families who remained behind experienced a

[1] *Ibid.* pp. 80–4. Orwin and Whetham, *op. cit.* pp. 227–39.

remarkable improvement in their conditions, real wages rising by about a fifth in the depression years. Now cries of 'rural depopulation' were heard, and the townsmen deplored the overcrowding and beating-down of urban wages for which the rural efflux was held responsible. Smallholdings Acts in 1892 and 1907 attempted to stem the flood by giving the labourer a personal interest in the land, but although there was a growth of smallholdings it was evident that the necessary capital and personal qualities for successful small-scale cultivation were rather rare, and indeed that the market for the produce of smallholdings was itself limited. More than Acts of Parliament were needed to turn a centuries-old rural proletariat into a race of peasant cultivators. The labourer remained a labourer, although he now had generally a cheap and somewhat better cottage, an allotment, education, better access to the amenities of towns, and after 1883 the vote.

Money wages still remained desperately low, below 15s. on average, and down to something less than 13s. in the low wage counties even in the first years of the twentieth century. Allowing for payments in kind the agricultural labourer received only two-thirds of the average industrial wage. It may be noted, however, that despite their low incomes and, too often, their miserable living conditions, the agricultural labourers were very nearly the most healthy manual workers in the country, having a death rate only little more than a quarter of that of general labourers, and only marginally higher than that of their employers, the farmers; indeed, the agricultural labourers had a higher expectation of life than had even the schoolmasters and civil servants. Even their low earnings could not offset the advantages for health of fresh air and a country environment.[1]

By 1911 the number of hired workers in agriculture in England and Wales had slowly risen again and stood at 665,000, but wages being what they were, the migration from the villages could not be checked, although since the 90's it had proceeded at a much reduced rate. The number of labourers now represented a proportion of only a little over three hired workers to each farmer or grazier, and the 60 per cent or so of farmers who employed labourers were still reducing their arable, using more machinery, cutting out inessential jobs and working harder themselves. Within 40 years after 1871 the proportion of the working population employed in agriculture had been nearly halved, and with only 7·6 per cent of the working population of 1911, farming had finally ceased to be the country's major industry.

[1] See E. H. Phelps Brown, *The Growth of British Industrial Relations* (1959) pp. 28–9, 35.

*References and suggestions for further reading:*

M. K. Ashby      *Joseph Ashby of Tysoe, 1859–1919: a study of English village Life* (1961).

*Joseph Arch: The Story of his Life, Told by Himself* (ed. the Countess of Warwick, 1898).

J. Caird      *English Agriculture in 1850–51* (2nd ed. with intro. by G. E. Mingay, 1968).

J. P. D. Dunbabin      The 'Revolt of the Field': The Agricultural Labourers' Movement in the 1870s,' *Past & Present* 26 (Nov. 1963).

T. W. Fletcher      'The Great Depression of English Agriculture 1873–1896' *Economic History Review* 2nd ser. XIII (1960–1).

Rider Haggard      *Rural England* (1902).

M. A. Havinden      *Estate Villages* (1966).

E. H. Hunt      *Regional Wage Variations in Britain 1850–1914* (1973).

E. L. Jones      'English Farming Before and During the Nineteenth Century' *Economic History Review* 2nd ser. XV (1962–3).

E. L. Jones      'The Agricultural Labour Market in England, 1793–1872' *Economic History Review*, 2nd ser. XVII (1964–5).

E. L. Jones      'The Changing Basis of English Agricultural Prosperity, 1853–73' *Agricultural History Review* X (1962).

Barbara Kerr      *Bound to the Soil: a Social History of Dorset 1750–1918* (1968).

R. Molland      'Agriculture *c.*1793–*c.*1870' *Victoria County History of Wiltshire* IV (ed. E. Critall, 1859).

C. S. Orwin and E. H. Whetham      *History of British Agriculture 1846–1914*, Ch. 4–8.

D. Spring      *The English Landed Estate in the Nineteenth Century* (1963).

R. Trow-Smith      *English Husbandry* (1951), Ch. X.

F. M. L. Thompson      *English Landed Society in the Nineteenth Century* (1963), Ch. 6, 9.

# Conclusion:
## The Agricultural
## Revolution and the Economy

The period from 1750 to 1880 with which we are concerned in this book confronted farmers with unprecedented demands not only for food, but also for raw materials and feed for draught animals, from a population which multiplied some five or six times; and in the course of responding to this new and revolutionary situation, the size, structure and technical efficiency of the industry were all greatly changed. Moreover, it was in this period that age-old trends in the countryside—the gathering of land into large estates, the movement to an optimum size of farms, and the growth of mixed grain-farming and animal husbandry—reached their climax. An agricultural system gradually built up since the Middle Ages came to maturity.

It was based, of course, upon the inherent qualities of the soil and the exigencies of climate and topography. England is one of the best farming countries in the world. Its varied soils and its relatively mild extremes of temperature and rainfall make possible the practice of mixed agriculture, i.e. animal and arable husbandry in combination, in most parts of the country in most months of the year; but favourable or unfavourable as the conditions imposed by nature might be, the human factor finally determined the rate and direction of the progress towards what we call the Agricultural Revolution; and in the case of England, as a whole, again owing to historical circumstances, the human factor was especially favourable to the development of progressive forms of agriculture from an early period. The social milieu in which English farming grew up encouraged a flexibility of response to the market which was denied farmers of many other lands, including most of the rest of the British Isles.

Even in Wales, where the system of landholding had been influenced by proximity to England and by intermarriage of Welsh and English families, the landlords consisted either of remote estate owners operating through agents or poverty stricken squires interested only in their rents and the exercise of their sporting rights; and it was not until the end of the eighteenth century that landlord investment in building and other improvements, with interest charges added to the rents, began to make any significant progress.

The institutional advantages enjoyed by English farmers were of two kinds: first, a landlord-tenant relationship involving certain conventional responsibilities on the part of the landlord in return for rent, a relationship which, because it was unknown in Ireland and Scotland was called, according to Thorold Rogers, 'the English system'. The second advantage English farmers enjoyed was a fluidity of movement between town and country which fertilized the countryside with the capital and commercial spirit of the urban business community and gave to the operations of farming and estate management a rationality of outlook which was rarely encountered elsewhere.

In regard to the landlord-tenant relationship, Thorold Rogers found that it involved, in the estates he had studied, not only the provision and maintenance of fixed capital in the form of farm buildings but the partial replacement of tenants' losses in bad years. The landlords in the cases mentioned were ecclesiastical corporations, New College and Magdalen College, and the time was the fifteenth century,[1] but the principle of landlord investment and even of some participation in the risks became generally recognized, however much it varied in actual practice; and alongside it ran, at least by the eighteenth century, a recognition of conventional security of tenure, without legal sanction, and dependent on the mutual understanding of landlord and tenant. As a factor in the progressive change in agricultural methods this could operate both ways, as we have shown in this book; but if a lazy tenant was encouraged in his laziness, an enterprising one had the necessary confidence to experiment even without legal security in the tenure of his holding; and it was the substantial tenant farmers and large owner-occupiers who took the lead in the technical innovations which contributed so essentially to the agricultural revolution.

[1] J. E. Thorold Rogers, *The Economic Interpretation of History* (1893) p. 169.

The second characteristic of mobility between town and country also has medieval antecedents. In the thirteenth century, London merchants bought country estates and founded landed families instead of merchant dynasties;[1] younger sons who were put to the law regarded land as the natural objective of a successful career, and for 300 years before the period dealt with by this book the English landed interest had been, to a significant extent, recruited from those who had reached the top in other walks of life.

Needless to say, many quickly fell back to the starting-point, but there was no lack of demand for the land that they released in their fall; and in spite of the wholesale pillage of the heritage of Crown and Church by the rising landed classes, the activity of the land market was determined more by a limitation of supply than of demand, and the resources of Ireland had to be called in to meet the insatiable appetite of those who were late in the queue or too poor to compete. Moreover, in the eighteenth century the sale of land was restricted by the growth of strict family settlements and also, to some extent, by developments in the capital market. The object of the former was to erect legal obstacles in the path of a reckless or panic-stricken heir who might try to meet his inherited debts by the sale of part or the whole of the family estates; and the result of the latter was to provide him with practical alternatives by making it easier for him to raise loans on the estate to meet his debts and to effect improvements. These greater facilities for borrowing had followed the fall in the rate of interest from 10 per cent in the early years of the seventeenth century to 6 per cent by 1680, and to 5 per cent or even 4 per cent by 1740, and lenders on long-term mortgage for these purposes and on these terms were never lacking.

The effect of these developments was to limit the power of the landowner to ruin his estate by reckless extravagance while at the same time making it easier for him, if he was so minded, to improve it by borrowing for productive investment; and though they strengthened the trend towards the consolidation of land ownership in a small patrician class, they never went so far as to stop up the channels by which new blood could find its way in to revivify the personnel of the landed interest and prevent it from becoming a closed caste.

The English landed interest had long accepted the idea that land was an agent of production and not merely an instrument of feudal and social prestige; and through their skilful manipulation of the landlord-tenant

[1] See Sylvia Thrupp, *The Merchant Class of Medieval London* (Chicago, 1948) Ch. VI.

system of management, through, particularly, their employment of able and progressive estate stewards or agents—men whose role in our agricultural development has not been sufficiently appreciated—the more astute landowners were able to get the best of both worlds, the world of political and social prestige which the mere possession of land in sufficient quantity conferred, and the world of commercial affluence which the careful selection and encouragement of good tenants could open. English agriculture, therefore, developed in a framework which permitted the early development of a farming system that could respond to market demand both in regard to the size of the farming unit and methods of cultivation; and as market demand fluctuated under the influence of price changes reflecting, for the most part, the changing impact of population pressure, it acquired a flexibility and a responsiveness that took it to the leadership of the world in farming practice.

The farmers themselves, as has been shown already, probably deserve the main credit for this achievement. Large landowners taken as a whole were more a class of consumers, borrowing heavily on the strength of their rents to meet the demands of luxurious living, but there can be no doubt that the number of genuine entrepreneurs among the landowning class was large enough to make a sizeable impact on developments both within and beyond the range of their immediate agricultural interests. Moreover, their possession of political power made them the ideal vehicle for schemes requiring Acts of Parliament, and they often found themselves involved in the promotion of local acts as a result of the importunities of traders and industrialists who wished to use them for this purpose. They were also, directly and indirectly, advancing their own interests. River navigations, turnpikes and canals were to a large extent aspects of agricultural expansion since they met the pressing need for improved transport of agricultural products, grain, malt, live cattle; lime, manure and, to an increasing extent, of bricks and other building materials for the immense rebuilding programme that was taking place in villages as well as in towns under the stimulus of expanding population and changing agriculture. The contribution of the humbler ranks of the rural community —the large body of small farmers and labourers who did the carting of iron and coal and building materials in slack periods of the year—should also be remembered. The development of industry was dependent, to a considerable extent, on the diversion of scarce resources, both human and material, from agriculture to forms of industrial enterprise.

In regard to the traffic in coal, its production and distribution was a major rural preoccupation in the coal-producing districts. The accumulation of coal at pit-heads during the winter months was a constant source of complaint both of producers in the country and consumers in the town and was perhaps the main single driving force behind the spate of Acts concerning transport that marked the general expansion from the late 1750s. Between 1750 and 1754 there were 93 Turnpike Acts, and as early as 1755 the first survey was made for a canal to join the Trent at Shardlow and the Weaver in Cheshire while a second was made in 1758 on behalf of Earl Gowan and Lord Anson. Of the 165 canal Acts passed between 1758 and 1801 no fewer than 90 were concerned with the transport of coal; and coal-owners were an integral part of the landed interest. The Duke of Bridgewater may have been the greatest of his class who sowed the seeds and reaped the harvest—in his case a fabulous one—of canal enterprise, but he had many imitators, and numerous other landed proprietors made themselves individually responsible for the promotion and completion of canals.[1] Of the £13,000,000 of canal company stock subscribed between 1758 and 1802, mainly from the localities which the projects were designed to serve, a very large proportion must have come from the pockets of landed proprietors, many of whom had local development in mind as much as—or more than—mere speculation. Such relatively unknown figures as Walter Spencer Stanhope, with extensive estates in Yorkshire and Derbyshire, interested in coal mines, iron manufacture, the cloth industry and promoter of the bill for the Barnsley Canal, and Joseph Wilkes of Overseal in Leicestershire, improving farmer, canal promoter, banker, cotton manufacturer and creator of the small industrial town of Measham, show the scope and versatility of the rural capitalist-turned-promoter; and there were many like them.[2] The railway age offers parallel cases of the transference of landed incomes and entrepreneurial leadership to non-rural objectives. Such names as Earl Fitzwilliam in Yorkshire; the Lowthers of Cumberland; Lord Durham and the Marquis of Londonderry; and perhaps particularly the Duke of Devonshire who literally called into being the town of Barrow-in-Furness, are reminders of the aristocratic origin of much of the industrial enterprise north of the Trent.

But the effects of the ebb and flow of agricultural incomes were not confined to the upper echelons of rural society. As Arthur Young pointed out in his *Travels in France*[3] one of the most important characteristics that distinguished the English countryside from that of its nearest

[1] See T. S. Ashton, *An Economic History of England: The Eighteenth Century* (1955) p. 74; and *Economic Fluctuations in England 1700–1800* (1960) pp. 5–6.

[2] See G. E. Mingay, *English Landed Society in the Eighteenth Century* (1963) Ch. 8.

[3] A. Young, *Travels in France 1787–9* (1892) pp. 132, 266.

neighbour and rival was the infinite gradation that linked the upper and the lower orders of the rural community. In France 'there are no gentle transitions from care to comfort, and from comfort to wealth. You pass at once from beggary to profusion.' Instead of lending money for improvement, the moneyed people in the country locked it up, and 'great cities have not the hundredth part of connection and communication with each other that much inferior places enjoy with us,' a sure proof of a lack of the 'consumption, activity and animation' that he found existing to such a marked degree in the English countryside.

'Consumption, activity and animation': these were certainly the characteristics of English rural society throughout the eighteenth century, and to large numbers in town and country, they frequently assumed a degree of urgency unknown to modern generations through the fluctuations of the harvest.[1] An unusually good harvest called for an increase in the army of harvesters—men, women and children who offered themselves, year by year, as to a national campaign that might in good years take on something of the glamour of a festival—to cut the grain, to bind, stook and haul it to the stackyard for the long winter tasks of threshing and carting to the mill or to market. As a result, in good years, wages rose while prices of provisions fell, though non-agricultural goods might become dearer because of the rising demand from labourers with a surplus to spend over their needs for food. Not only the labouring classes, but tradesmen, manufacturers, middlemen who dealt in or used agricultural products, millers, bakers, starch-makers, stationers, bookbinders, linen-printers, trunk-makers, paper-hangers, benefited from the low price of flour; distillers, brewers, maltsters, innkeepers enjoyed a good barley crop; makers of shoes and harness, soap-boilers, candle-makers, cutlers, glue manufacturers were better off when cattle and sheep were plentiful and cheap; spinners and weavers of wool had a longer and more lucrative harvest holiday, or sold their labour to manufacturers at a higher price; everybody had more to spend on wants other than food. That is, everybody but farmers and landlords, who grumbled when prices were low and wages tended to rise.

Similarly, when harvests fell short, there was a wave-like motion of falling incomes, and rising unemployment side by side with rising food prices, culminating in bad years in riots and attacks on millers, bakers and butchers. At the same time, the incomes of farmers—especially the larger ones—rose since they could sell their surplus at a disproportion-

[1] For a full analysis of the intricate problems posed by harvest fluctuations, see T. S. Ashton, *Economic Fluctuations in England, 1700–1800* Ch. 2.

ately high price owing to the inelasticity of supply and demand of agriculture produce, especially grain. Arthur Young spoke of a 'universal circulation of intelligence, which in England transmits the least vibration of alarm, with electric sensitivity, from one end of the Kingdom to another.' He was speaking in political terms, but the channels of sensitivity had been worn smooth and swift by the hopes and fears aroused by harvest fluctuations among a population whose daily bread might literally depend upon them.

The part played by the 'consumption, activity and animation' engendered by the ceaseless round of English agriculture has given rise to a discussion of great interest and importance to those concerned with the timing of the Industrial Revolution. Professor A. H. John[1] has argued that the low prices of the first half of the century raised real wages and enhanced living standards of all classes; a stimulus was imparted to the home market on such a scale that the acceleration of population in the second half of the century was a prelude to a general expansion that culminated in the Industrial Revolution. Other students of the problem[2] have seen in this long period of low prices a depressing effect upon investment and a deceleration in the rate of expansion owing to the relative fall of incomes of farmers and landlords, especially in the second quarter of the eighteenth century; and have found the springboard for the upward thrust in the next quarter in the simultaneous rise of agricultural prices (initiated by the heavy exports of grain in the 1750s) and of population from the middle of the century, a combined movement which restored prosperity to farmers and landlords and encouraged a boom in agricultural investment, building and transport, with favourable repercussions on related industries.[3]

The rising standard of life to which the labouring classes had become accustomed no doubt contributed to higher output by providing them with an incentive to increase their efforts to maintain it as prices rose, and also to enjoy the lengthening schedule of demand made possible by the

---

[1] See, e.g., 'Aspects of English Economic Growth in the first half of the Eighteenth Century', in *Essays in Economic History*, ed. E. Carus-Wilson, II (1962) pp. 365–7.

[2] See Deane and Cole, *op. cit.* pp. 92–7.

[3] The parallel trends may be illustrated as follows:

| Approximate | 1720/30 | 1730/40 | 1740/50 | 1750/60 | 1760/70 | 1770/80 |
|---|---|---|---|---|---|---|
| % rise of population | −0·7 | +·08 | +4·0 | +6·6 | +6·8 | +6·4 |
| Wheat Prices (1700/1 = 100) | 99 | 84 | 84 | 101 | 117 | 136 |
| Enclosure Acts | 25 | 39 | 36 | 137 | 385 | 660 |

increasing supply of better houses, clothes, household goods and of agricultural products, particularly wheaten bread, meat and dairy produce. As Professor John has said, 'The buoyancy of the domestic market . . . [had] provided a favourable environment for the introduction of new kinds of goods . . . cheap crockery, japanned wares, lace, Sheffield plate . . . cheap mixed fabrics . . . [It] encouraged manufacturers to be inventive and to direct production toward cheapness, as fashion ceased to be a prerogative of the rich.'[1]

The effect upon agriculture of the upward turn of prices was slow to reveal itself. The demand for grain was cushioned by the export trade and the consumption of grain for alcohol. Both could be drawn upon for the home supply of bread grains, and the first phase of agricultural expansion seems to have been aimed at raising the supply of wool and leather, that is for the products of pasture rather than of arable farming. The consumption of spirits fell after the Gin Act of 1751, and it was not until the poor harvests between 1764 and 1775 that the export surplus in grain was turned to a deficit. This sudden reversal in the position of supply and demand was all the harder to bear owing to the increased consumption of wheaten bread as well as meat and dairy produce which the long period of low prices had made possible; and the Agricultural Revolution was speeded on its way by extensive and serious food riots as prices rose, especially in the last years of the century when the rising pressure of demand, already reaching explosive force, was reinforced by the exigencies of war.

The whole national effort now turned upon the ability of the agricultural community to meet the unprecedented demand not only for food but also for raw materials, especially wool, timber and leather, and also for fodder for the army of horses that was needed to keep the traffic moving on roads, rivers and canals. Imports now began to play a minor but important role: the expanded supply of both meat and corn included to an increasing extent produce from Ireland, and by the end of the century British imports of meat and dairy produce totalled £2.1 million compared to an import of £2.6 million for corn. Relatively poor harvests in the last quarter of the century, together with the unprecedented rise of population, caused demand for agricultural goods to rise faster than supply, and Britain began to move into the position of a permanent importer of grain. Imports increased during the French Wars, but in the years of harvest failure, prices rose to famine heights: 1795, 1799–1800, 1808 to

[1] A. H. John, 'Agricultural Productivity and Economic Growth in England, 1700–1760' *Journal of Economic History* XXV (1965) pp. 24–5.

1812 were crisis years, and the poor suffered very real hardship. The agricultural industry responded by taking in new land rather than by developing new practices, and between 1793 and 1815 about 1,000,000 acres of common pasture and waste were enclosed mainly for arable. There was a great ploughing up of old enclosed pastures, and the value of the total product rose, it has been calculated, from £75·5 million in 1801 to £107·5 million in 1811.

Prices fell after the war, but expansion continued especially on light soils; a further 200,000 acres of common pasture and waste were enclosed between 1815 and 1845, and low prices provided the spur that was necessary to start a new phase of agricultural improvement. Many farmers who were unable or unwilling to take advantage of the more advanced methods or were too closely tied to the market in cereals found themselves in difficulties. This was especially true of the farmers on the heavy soils—the midland clays and Keuper marls—the traditional corn lands of England. Farmers in these regions found themselves outclassed and undersold by the farmers in the light soil regions where mixed farming permitted greater variety of products and higher yields at lower costs. The clay soil farmers were soon in great distress: they were faced with increased imports from abroad and with deflationary policies at home which lowered prices still further. They clamoured for protection in the vain hope that by raising the price of imported goods the general level of prices in the home market would be raised to profitable levels. In this they were disappointed owing to the competition not of the foreigners but of their more favourably situated neighbours. What was needed—and what was accomplished—was a wholesale readjustment of clay soil farming by means of a large-scale conversion to pasture, a heavy investment in drainage and the elimination of weaker farmers. At the same time, the development of farm machinery—a technological revolution, in fact—the growing impact of the agricultural sciences and the coming of the railways were preparing the way for the final phase of the Agricultural Revolution, the era of high farming.

By this time, however, the specifically English character of the transition to modern scientific agriculture and the mass-production of food had run its course, and influences, both European and American, began to assert a leadership in fields for which they were specially adapted. British agriculture now had to take its place in a world-wide pattern of supply and demand which it had done much to create but which it could do nothing to control. The Agricultural Revolution, however, had performed its role in the process of industrialization. Output had risen almost as fast as population, and as late as 1868 it was estimated that no less than 80 per cent of the food consumed in the United Kingdom by a highly urbanized and industrialized population had been grown at home. By maintaining its impetus, agriculture had enabled industrialization to advance to the point at which the country could draw upon the food surplus of its customers in all parts of the world. Other sources of wealth had been tapped, and agriculture, which at the beginning of the century provided 40 per cent of the national product, by 1880 provided only 10 per cent. At the same time, although the actual size of the labour force in farming remained practically the same in 1881 as in 1801, the proportion of the national labour force employed in agriculture had fallen from 35·9 per cent to 12·6 per cent,[1] and the share of national capital from more than a half to less than one-fifth.

An economic transformation on this scale could not fail to have profound repercussions on the balance of social and political forces, and the fact that this readjustment was carried through peacefully and, indeed, under the direction of the leaders of a social class now entering upon its decline, is one of the most remarkable political phenomena of the age. The historian of this transition has explained it partly on the grounds that the process of absorption into the landed interest from other classes 'must be accounted a prime reason for the failure of the cleavage between capitalists and landowners to become so deep as to be unbridgeable'.[2] There was also a degree of wisdom—and difficult as it is to believe—a deference to public opinion, that help to explain how a proud and opulent aristocracy could take a peaceful and indeed constructive part in dismantling their own structure of political power. There were some who thought in terms of violent resistance like the Duke of Buckingham, who took the cannon from his yacht to resist the Reform of 1832, or the Duke of Newcastle, who fortified his house at Clumber, but the leaders fought the battle in terms of party politics and placed party advantage over class prejudices. Hence it came about that a parliament of landowners passed the franchise Acts of 1832 and 1867, imposed agricultural protection in

[1] The figures and proportions for the whole period are given by Deane and Cole (*op. cit.* pp. 142–3) as follows:

|  | 1801 | 1811 | 1821 | 1831 | 1841 | 1851 | 1861 | 1871 | 1881 |
|---|---|---|---|---|---|---|---|---|---|
| No. of persons in agriculture, forestry and fishing | 1·7m | 1·8m | 1·8m | 1·8m | 1·9m | 2·1m | 2·0m | 1·8m | 1·7m |
| Percentage of labour force | 35·9 | 33·0 | 28·4 | 24·6 | 22·2 | 21·7 | 18·7 | 15·1 | 12·6 |

[2] F. M. L. Thompson, *English Landed Society in the Nineteenth Century* (1963) p. 22.

1815, and abolished it in 1846. Their political eclipse was inevitable, but that they should have presided over it themselves is perhaps the most successful manœuvre in their remarkably successful career.

The end of the Agricultural Revolution is also the end of an epoch in English farming. As farmers and landlords faced the consequences of free trade which they were powerless to stem by political action, they had to meet a new type of challenge. Farmers' production became increasingly influenced by the quality and prices of imported food, and English farmers had to move away from the old corn staple towards commodities such as meat, milk and vegetables, which were less affected by imports. The growth of international trade thus brought about the new era for English farming which the repeal of 1846 had prematurely announced.

The rapid increase in imports beginning in the 1870s accelerated the pace of change in English farming and gave rise to a new structure of production, landownership and farm tenure. Wheat and other grains rose by 90 per cent between 1875 and 1900; butter and cheese by 110 per cent, meat imports by 300 per cent. These figures, however, do not indicate the relative impact of the increased imports in the several branches of agriculture, because while wheat prices fell by half in the years after 1875, the prices of home-produced butter, cheese and meat all fell less quickly and in the end by little more than 10 per cent. These very uneven movements in prices resulted from two aspects of the market situation: first, that wheat imports were already heavy before 1875, while imports of dairy produce and meat were at that time a relatively small part of the total supply; consequently the percentage increases in imports represented quite different orders of magnitude, and even in 1900 English pasture farmers still commanded a large share of the home market for their products. The second feature of the market was the growing shift of consumers' demand away from bread and towards meat and dairy produce, a consequence of rising real incomes. During the 'great depression' period the population increased by about 10,000,000. The market for food was thus expanding rapidly, but because consumers were now eating less bread and far more meat and dairy produce (and also more vegetables and fruit) the market for pasture products was growing more rapidly than that for wheat. The consequence was that wheat acreage fell sharply, while the output of cattle, pigs and farm horses, and of oats, fruit and vegetables all rose. There was, therefore, no *general* depression in English farming.

Agriculture was still a major industry but a relatively declining one. As late as 1851 it continued to employ over a fifth of the occupied population, equal to more than half of the population engaged in manufacturing industries. By 1901 it employed less than one-tenth of the occupied population. Its share of the national income fell further still, from 20·3 per cent in 1851 to 6·4 per cent in 1901, although in money terms it remained remarkably stable—about £105,000,000. Indeed, it was owing to the great expansion of British farming and of home-grown food supplies that industry until this time had been able to keep up its rate of growth; but this role was now ending with the expansion of world-wide trade and the exchange of manufactured goods for foreign food surpluses. By virtue of its own success, agriculture was forced to adapt itself to a new economic climate and to accept a humbler role in the economy. Arable farmers were faced with severe problems, and many found them insoluble. Landlords were also among the sufferers, especially those in arable areas who were receiving little income from their estates in spite of past and current investment. As a class, the landlords had also to step down from the pinnacle of power they had occupied for nearly 300 years and even to see their leaders engaged in the delicate task of grafting the new industrial democracy on to the old plant of the hereditary aristocracy. It is a remarkable spectacle; and perhaps it may be said that in all their long and successful tenure of the monopoly of political power, there is nothing becomes them more than the manner of their leaving it.

*References and suggestions for further reading:*

| | |
|---|---|
| T. S. Ashton | *An Economic History of England: the Eighteenth Century* (1955). |
| T. S. Ashton | *Economic Fluctuations in England 1700–1800* (Oxford, 1959), Ch. 2. |
| J. D. Chambers | *The Vale of Trent* (Supplement No. 3 to the Economic History Review), Ch. IV. |
| J. D. Chambers | *The Workshop of the World* (1961), Ch. 1, 3. |
| Phyllis Deane and A. W. Cole | *British Economic Growth 1688–1959* (Cambridge, 1962), Ch. II (3). |
| Phyllis Deane and H. J. Habakkuk | 'The Take-off in Britain', in *The Economics of Take-off into Sustained Growth* (ed. W. W. Rostow, 1963). |
| A. H. John | 'Agricultural Productivity and Economic Growth in England 1700–1760', *Journal of Economic History* XXV (1965). |

*References and suggestions for further reading:*

A. H. John — 'Aspects of English Economic Growth in the First Half of the Eighteenth Century', in *Essays in Economic History* (ed. E. M. Carus-Wilson), II (1962).

E. L. Jones — *Agriculture and Economic Growth in England 1650–1815* (1967).

G. E. Mingay — *English Landed Society in the Eighteenth Century* (1963), Ch. 8.

P. J. Perry — *British Farming in the Great Depression 1870–1914* (1974).

J. E. Thorold Rogers — *The Economic Interpretation of History* (1893), Ch. VIII.

Charles Wilson — *England's Apprenticeship 1603–1763* (1965), Ch. 2, 7.

*The sources for the figures are as follows:*

Fig. 1   *B.P.P.* VIII pt. 2 (1836), p. 501; B. R. Mitchell and Phyllis Deane, *Abstract of British Historical Statistics* (Cambridge, 1962), pp. 455, 488.

Fig. 2   Mitchell and Deane, *op. cit.*, pp. 487–9; G. R. Porter, *Progress of the Nation* (1851), p. 597.

Fig. 3   J. Caird, *English Agriculture 1850–51* (1852), frontispiece.

Fig. 4   F. M. L. Thompson, *English Landed Society in the Nineteenth Century* (1963), pp. 218–20, 231–5; R. J. Thompson, 'An Inquiry into the Rent of Agricultural Land in England and Wales in the Nineteenth Century' *Journal of the Royal Statistical Society* LXX (1907); H. A. Rhee, *The Rent of Agricultural Land in England and Wales 1870–1943* (1949), pp. 44–6; and private sources.

Fig. 5   Mitchell and Deane, *op. cit.*, pp. 474, 488–9.

# Index

# Index

Acreage, arable, 183
Agricultural Revolution, 3–5, 10, 13–14, 199, 206–10
Allom, Joseph, 66
Allotments, 86, 97–8, 101–2, 134–5, 191, 192
Alternate husbandry, 4, 54, 56, 61–2, 98; on Norfolk estates, 59
Andover, 22
Anglesey, 30, 31
Animal disease, 19, 38, 42, 171, 179, 182, 186. *See also* Cattle plague, Liver rot
Animal husbandry, 13; output of, 35
Animal products, prices of, 109, 184, 209
Anson, Lord, 203
Anti-Corn Law League, 152–6, 161
Anti-League, 154–5
Arbuthnot, John, 57, 70
Arch, Joseph, 195–6, 198
Argyle, 185
Artificial grasses, 8, 54–8, 60; in open fields, 51–2
Ashby, Joseph, 192, 198
Ashby, M. K., 192n, 198
Ashton, T. S., 12n, 33, 53, 82, 203n, 204n, 210
Ashworth, W., 162n
Axholme, Isle of, 137
Aylesbury, Vale of, 22, 107

Bacon, 23, 107, 180–1
Baker, Robert, 154–5
Bakewell, Robert, 4, 13, 36, 39, 61, 66–9, 74
Barley, 16, 56, 62, 107–8, 159, 204; prices of, 108, 113, 127, 159, 178
Barnes, D. G., 109n, 122n, 123n, 124, 127n, 147, 154n, 160n, 169
Barrett, Elizabeth, 194
Barrow-in-Furness, 203
Basic slag, 170
Barnet Fair, 30

Batchelor, T., 96
Beccles, 22
Bective, Lord, 173
Bedford, 22
Bedford, Dukes of, 11, 101, 136, 192
Bedfordshire, 22, 23, 28, 29, 37, 96, 103, 130, 136
Beecham, H. A., 49n
Beef, 107–8; output of, 35–6; prices of, 111, 113
Beighton (Derbys.), 51
Bell, Patrick, 72
Bellerby, J. R., 182n
Belvoir, Vale of, 57
Bennett, William, 130
Beresford, M., 9n
Berkeley, Vale of, 94
Berkshire, 23, 73, 103, 172
Black Country, 25, 27
Black Death, 5–6
Blaug, M., 119n, 120n, 134n, 141n
Board of Agriculture, 48, 57, 74, 85, 118, 121–2, 129, 135; *Reports* of, 73–4, 96, 121
Brecknockshire, 30
Bridgwater, Duke of, 203
Bright, John, 153, 154, 155
Bristol, 22, 26, 27, 77
Bromley (Kent), 90
Brougham, Lord, 126–7, 135
Brown, E. H. Phelps, 197n
Brown, Lucy, 151n, 169
Buckingham, Duke of, 208
Buckinghamshire, 23, 27, 41, 49, 94, 96, 195
Buildings, expenditure on, 131, 163, 175–7
Burford Fair, 29
Butter, 16, 25, 107–8, 125, 182, 186; imports of, 171, 180–1, 209
Buxton, 68
Byron, Lord, 124